Microwave Circuits for 24 GHz Automotive Radar in Silicon-based Technologies

Vadim Issakov

Microwave Circuits for 24 GHz Automotive Radar in Silicon-based Technologies

Vadim Issakov
Infineon Technologies AG
Am Campeon 1-12
85579 Neubiberg
Germany
vadim.issakov@infineon.com

Title of the Dissertation EIM-E/267, University of Paderborn, Germany, 2010: Microwave
Circuits for 24GHz Radar Front-End Applications in CMOS and Bipolar Technologies

ISBN 978-3-642-13597-2   e-ISBN 978-3-642-13598-9
DOI 10.1007/978-3-642-13598-9
Springer Heidelberg Dordrecht London New York

Library of Congress Control Number: 2010932572

© Springer-Verlag Berlin Heidelberg 2010
This work is subject to copyright. All rights are reserved, whether the whole or part of the material is
concerned, specifically the rights of translation, reprinting, reuse of illustrations, recitation, broadcasting,
reproduction on microfilm or in any other way, and storage in data banks. Duplication of this publication
or parts thereof is permitted only under the provisions of the German Copyright Law of September 9, 1965,
in its current version, and permission for use must always be obtained from Springer. Violations are
liable to prosecution under the German Copyright Law.
The use of general descriptive names, registered names, trademarks, etc. in this publication does not
imply, even in the absence of a specific statement, that such names are exempt from the relevant protective
laws and regulations and therefore free for general use.

Cover design: WMXDesign GmbH

Printed on acid-free paper

Springer is part of Springer Science+Business Media (www.springer.com)

Preface

There are continuous efforts focussed on improving road traffic safety worldwide. Numerous vehicle safety features have been invented and standardized over the past decades. Particularly interesting are the driver assistance systems, since these can considerably reduce the number of accidents by supporting drivers' perception of their surroundings. Many driver assistance features rely on radar-based sensors. Nowadays the commercially available automotive front-end sensors are comprised of discrete components, thus making the radar modules highly-priced and suitable for integration only in premium class vehicles. Realization of low-cost radar front-end circuits would enable their implementation in inexpensive economy cars, considerably contributing to traffic safety.

Cost reduction requires high-level integration of the microwave front-end circuitry, specifically analog and digital circuit blocks co-located on a single chip. Recent developments of silicon-based technologies, e.g. CMOS and SiGe:C bipolar, make them suitable for realization of microwave sensors. Additionally, these technologies offer the necessary integration capability. However, the required output power and temperature stability, necessary for automotive radar sensor products, have not yet been achieved in standard digital CMOS technologies. On the other hand, SiGe bipolar technology offers excellent high-frequency characteristics and necessary output power for automotive applications, but has lower potential for realization of digital blocks than CMOS.

This work presents the design, implementation, and characterization of microwave receiver circuits in CMOS and SiGe bipolar technologies. The applicability of a standard digital 0.13 μm CMOS technology for realization of a 24 GHz narrowband radar front-end sensor is investigated. The unlicensed industrial, scientific and medical (ISM) frequency band at 24 GHz is particularly interesting for radar applications, due to its worldwide availability and the possibility of inexpensive packaging in this frequency range.

The low-noise amplifier (LNA) and mixer receiver building blocks have been designed in CMOS and bipolar technologies. These building blocks have been integrated into receiver and transceiver front-ends. The performance stability of the circuits is compared over a very wide temperature range from -40 to 125 °C. Additionally, ESD protection techniques are considered. Further, advanced modeling and de-embedding techniques, required for accurate circuit characterization, are investigated. The presented circuits are suitable for automotive, industrial and consumer applications, as e.g. lane-change assistant, door openers or alarms.

This manuscript is based on the dissertation entitled "Microwave Circuits for 24 GHz Radar Front-End Applications in CMOS and Bipolar Technologies" submitted to the University of Paderborn. The research work was supported under the German BMBF funded project EMCpack/FASMZS 16SV3295 and was carried out in close collaboration with Infineon Technologies AG, Neubiberg, Germany.

I would like to express the deepest gratitude to my advisor Prof. Dr.-Ing. Andreas Thiede for his kind guidance, support, patience and insight throughout my research at the University of Paderborn. His valuable advice and inspiring ideas have advanced my work and encouraged me to research deeper. I highly appreciate his great efforts, amiable attention and understanding evinced in the guidance of my research work.

Furthermore, my debt of gratitude is owed to Prof. Dr.-Ing. Andreas Thiede and Prof. Dr.-Ing. Dr.-Ing. habil. Robert Weigel for reviewing this manuscript.

In addition, I would like to express my sincere appreciation to Dr. Werner Simbürger for enabling and supporting my activities at Infineon Technologies AG, Neubiberg, Germany. His sustained encouragement and valuable discussions have contributed a great deal to this work.

A very special thank you goes to Dr. Herbert Knapp and Dr. Marc Tiebout of Infineon Technologies AG for many valuable discussions, suggestions and their continuous support throughout the research. Thanks also goes to Maciej Wojnowski of Infineon Technologies AG for the kind support with on-wafer measurements, packaging and numerous interesting discussions about de-embedding and calibration techniques. My thanks also go to Mirjana Rest for the initial support with the layouts and job deck viewing. Furthermore, I would like to thank my Infineon colleagues Dr. Ronald Thüringer, Dr. Winfried Bakalski, Dr. Ludger Verweyen, Domagoj Šiprak, Yiqun Cao, David Johnsson and Kevni Büyüktas for their kind support.

A kind thank you goes to Dr. Volker Winkler of EADS, Ulm, Germany for his valuable help with measurements and radar system aspects. Additionally, I would like to thank the colleagues Dr. Linus Maurer, Günter Haider and Shoujun Yang from Danube Integrated Circuit Engineering (DICE) GmbH, Linz, Austria for helpful comments and supporting this work.

I wish to express my sincere appreciation to efforts of Mr. Peter Jupp of Peak RF Ltd., Cambridge, UK for carefully reading through this manuscript and refining the English grammar in this work.

I would like to thank my fiancee Elisabeth Hofmann for her support and patience. As well, I express my sincere gratitude to my parents Eduard and Maya Issakov for the continuous encouragement, motivation, care and their priceless support.

Vadim Issakov
Munich, Germany
May 2010

Contents

1	**Introduction**	1
	References	4
2	**Radar Systems**	5
	2.1 Radar Principle	5
	2.2 Radar Equation and System Considerations	6
	2.3 CW and Frequency-Modulated Radar	8
	2.3.1 Doppler Radar	8
	2.3.2 Frequency-Modulated Radar	9
	2.3.2.1 Linear FM Continuous-Wave Radar	9
	2.4 Angle Detection	11
	2.5 Frequency Regulations	12
	2.6 Receiver Architectures	14
	2.6.1 Homodyne	14
	2.6.2 Heterodyne	15
	2.7 Status of Automotive Radar Systems	16
	2.8 Technology Requirements for Radar Chipset	17
	References	17
3	**CMOS and Bipolar Technologies**	19
	3.1 CMOS Technology	19
	3.1.1 MOSFET Layout and Modeling Considerations	20
	3.1.2 Devices Available in C11N	22
	3.2 Bipolar Transistors	23
	3.2.1 HBT Layout and Modeling Considerations	24
	3.2.2 Devices Available in B7HF200	25
	3.3 Technology Comparison	26
	3.3.1 Transistor Performance	26
	3.3.2 Metallization and Passive Components	29
	References	31

viii   Contents

4   Modeling Techniques 33
   4.1   Analytical Fitting of On-Chip Inductors 33
      4.1.1   Series Branch Parameters Fitting 36
      4.1.2   Shunt Branches Parameters Fitting..................... 38
      4.1.3   Results Verification 40
   4.2   Transistor Finger Capacitance Estimation 42
   References 45

5   Measurement Techniques 47
   5.1   S-parameter De-embedding Techniques 48
      5.1.1   Extension of Thru Technique for De-embedding of
             Asymmetrical Error Networks 49
            5.1.1.1   Theory 49
            5.1.1.2   Result Verification 52
      5.1.2   De-embedding of Differential Devices using
             cascade-based Two-Port Techniques 54
            5.1.2.1   Theory 54
            5.1.2.2   Result Verification 60
   5.2   Differential Measurements using Baluns 63
      5.2.1   Theoretical Analysis 64
            5.2.1.1   Back-to-Back Measurement 65
            5.2.1.2   DUT Measurement.......................... 67
            5.2.1.3   Insertion Loss De-embedding Error 68
      5.2.2   Measurement Verification 69
   References 74

6   Radar Receiver Circuits 77
   6.1   Low-Noise Amplifiers 78
      6.1.1   LNA in CMOS Technology 78
      6.1.2   LNA in SiGe:C Technology 83
      6.1.3   Measurements of CMOS and SiGe LNAs 86
      6.1.4   LNA Results Summary and Comparison 91
   6.2   Mixers... 92
      6.2.1   Active Mixers 93
            6.2.1.1   Active Mixer in CMOS Technology 93
            6.2.1.2   Active Mixer in SiGe Technology 95
            6.2.1.3   Measurements of CMOS and SiGe Active Mixers . 97
            6.2.1.4   Active Mixers Results Summary and Comparison . 101
      6.2.2   Passive Mixers 102
            6.2.2.1   Passive Resistive Ring Mixer in CMOS
                   Technology 102
            6.2.2.2   Passive Bipolar Mixer in SiGe Technology 105
            6.2.2.3   Measurements of CMOS and SiGe Passive Mixers 107
            6.2.2.4   Passive Mixers Results Summary and Comparison 110
      6.2.3   Comparison of Active and Passive Mixers 111

Contents

ix

6.3	Single-Channel Receivers		112
	6.3.1	Design of Active and Passive Receivers in CMOS	113
	6.3.2	Receiver Measurements and Analysis	113
		6.3.2.1 Chip Size	114
		6.3.2.2 Power Consumption, Gain and Noise Figure	114
		6.3.2.3 Linearity	116
		6.3.2.4 Required LO Power	118
		6.3.2.5 Isolation	119
		6.3.2.6 Temperature Performance	120
	6.3.3	Receiver Results Summary and Comparison	121
6.4	IQ Receivers		122
	6.4.1	Design of IQ Receivers	122
		6.4.1.1 IQ Receiver in CMOS Technology	122
		6.4.1.2 IQ Receiver in SiGe Technology	124
	6.4.2	IQ Receiver Measurements	125
	6.4.3	IQ Receiver Results Summary and Comparison	131
6.5	Integrated Passive Circuits		132
	6.5.1	Circuit Design and Layout Considerations	132
		6.5.1.1 On-Chip $180°$ Power Splitter/Combiner	132
		6.5.1.2 On-Chip $90°$ Power Splitter/Combiner	134
		6.5.1.3 On-Chip $180°$ Hybrid Ring Coupler	136
	6.5.2	Realization and Measurement Results	137
		6.5.2.1 On-Chip $180°$ Power Splitter/Combiner	137
		6.5.2.2 On-Chip $90°$ Power Splitter/Combiner	138
		6.5.2.3 On-Chip $180°$ Hybrid Ring Coupler	140
	6.5.3	Results Summary and Discussion	143
6.6	Circuit-Level RF ESD Protection		144
	6.6.1	Overview of Circuit-Level Protection Techniques	145
	6.6.2	Virtual Ground Concept	147
		6.6.2.1 Concept Verification by Circuit Simulation	149
		6.6.2.2 Concept Verification by HBM Measurement	150
		6.6.2.3 Concept Verification by TLP Measurement	151
	6.6.3	Transformer Protection Concept	153
		6.6.3.1 Test LNA Circuit Design	155
		6.6.3.2 Test LNA Realization and Measurement	156
		6.6.3.3 Concept Verification by TLP Measurement	157
	References		158
7	**Radar Transceiver Circuits**		165
7.1	IQ Transceiver in CMOS		166
	7.1.1	IQ Transceiver Circuit Design	166
	7.1.2	Measurements of Transceiver	169
	7.1.3	Results Summary and Comparison	171
7.2	Merged Power-Amplifier-Mixer Transceiver		173
	7.2.1	System Considerations	173

	7.2.2	Power-Amplifier-Mixer Circuit Design	174
	7.2.3	PAMIX Measurements	176
	7.2.4	Results Summary and Comparison	179
	References		180

8 Conclusions and Outlook 181

A LFMCW Radar 185
References 188

B FSCW Radar 189
References 190

C Surface Charge Method 191
C.1 Surface Charge Method Theory 191
C.2 Meshing of the Multifinger Layout 194

D Measurement of Active Circuits 197
D.1 Measurement Techniques 197
D.2 LNA Characterization 200
    D.2.1 S-parameter Measurement 200
    D.2.2 Noise Figure Measurement 200
    D.2.3 Linearity Measurement 202
D.3 Mixer and Receiver Characterization 203
    D.3.1 Conversion Gain Measurement 203
    D.3.2 Noise Figure Measurement 203
    D.3.3 Linearity Measurement 204
References 205

Index 207

Acronyms

ABS	Anti-lock braking system
AC	Alternating current
ACC	Adaptive cruise control
ADAC	Der Allgemeine Deutsche Automobil Club (German automobile club)
ADAS	Advanced driver assistance systems
ADC	Analog to digital converter
ADS	Advanced Design System
B7HF200	Infineon's 0.35 µm SiGe:C bipolar technology
Balun	Balanced-to-unbalanced converter
BEC	Base emitter collector
BEOL	Back end of line
BGA	Ball Grid Array
BiCMOS	Bipolar CMOS
BJT	Bipolar junction transistor
BOM	Bill of materials
C11N	Infineon's 0.13 µm CMOS technology
CAD	Computer aided design
CB	Common-base
CCD	Charge-coupled device
CDM	Charged device model
CE	Common-emitter
CG	Common-gate
CML	Current-mode logic
CMOS	Complementary metal-oxide-semiconductor
CPI	Coherent processing interval
CS	Common-source
CW	Continuous wave
DC	Direct current
DIBL	Drain-induced barrier lowering
DR	Dynamic range
DSB	Double sideband

DTI	Deep trench isolation
DTSCR	Diode-triggered silicon-controlled rectifier
DUT	Device under test
ECC	Electronic Communications Committee
EIRP	Equivalent isotropically radiated power
EM	Electromagnetic
ENR	Excess noise ratio
ESD	Electrostatic discharge
ESP	Electronic stability programme
ETSI	European Telecommunications Standards Institute
EU	European Union
FEOL	Front end of line
FET	Field-effect transistor
FFT	Fast Fourier transform
FMCW	Frequency-modulated continuous-wave
FSCW	Frequency-stepped continuous-wave
FSK	Frequency-shift keying
GaAs	Gallium-arsenide
HBM	Human body model
HBT	Heterojunction bipolar transistor
HF	High-frequency
HS	High-speed
HV	High-voltage
IC	Integrated circuit
IF	Intermediate frequency
IFA	Intermediate frequency amplifier
IIP3	Input-referred third-order intercept point
IMFDR	Intermodulation free dynamic range
InP	Indium-phosphide
IP1dB	Input-referred 1dB compression point
IP3	Third-order intercept point
IQ	In-Phase / Quadrature
ISM	Industrial, scientific and medical
LDD	Lightly doped drain
LFM	Linear frequency modulation
Lidar	Light detection and ranging
LNA	Low-noise amplifier
LO	Local oscillator
LRR	Long-range radar
LRRM	Line-Reflect-Reflect-Match
LSE	Least-square error
MIM	Metal-insulator-metal
MM	Machine model
MOS	Metal-oxide-semiconductor
MOSFET	Metal-oxide-semiconductor field-effect transistor

MRR	Mid-range radar
NF	Noise figure
NFM	Noise figure meter
NLVT	Low threshold voltage NMOSFET
NMOSFET	n-channel MOSFET
OIP3	Output-referred third-order intercept point
OP1dB	Output-referred 1dB compression point
P1dB	1dB compression point
PA	Power amplifier
PCB	Printed circuit board
PLL	Phase-locked loop
PLVT	Low threshold voltage PMOSFET
PMOSFET	p-channel MOSFET
PTAT	Proportional to absolute temperature
Radar	Radio detection and ranging
RCS	Radar cross section
RF	Radio frequency
SB	Single-balanced
SCM	Surface charge method
SCR	Silicon-controlled rectifier
SDM	Socketed device model
SE	Single-ended
SFDR	Spurious free dynamic range
SiGe	Silicon-germanium
SiO_2	Silicon-dioxide
SiP	System in package
SNR	Signal to noise ratio
SoC	System on chip
SOLT	Short-Open-Load-Thru
Sonar	Sound navigation and ranging
SPA	Spectrum analyzer
SRF	Self-resonance frequency
SRR	Short-range radar
SSB	Single sideband
STC	Sensitivity time control
STI	Shallow trench isolation
T/R	Transmit/receive
TaN	Tantalum-nitride
TL	Thru-Line
TLP	Transmission line pulse
TRL	Thru-Reflect-Line
TSLP	Thin Small Leadless Package
UHS	Ultra-high-speed
UWB	Ultra-wideband
VCO	Voltage-controlled oscillator

| VNA | Vector network analyzer |
| VQFP | Very small Quad Flat Package |

Chapter 1
Introduction

Increasing road traffic safety is a major objective of governments across the world. In particular, the European Union (EU) has set a challenging objective of halving the number of road accident victims by 2010 [1]. Active on-board safety features offer an approach with a high potential for achieving this target. It has been observed over the past decades that the decrease in the number of victims is related to technological innovations of the automotive safety, such as seatbelts, anti-lock braking system (ABS), airbags or electronic stability programme (ESP), as shown in Fig. 1.1 (*data source*: ADAC). Future generations of active safety equipment will be based on the advanced driver assistance systems (ADAS) including e.g. adaptive-cruise control (ACC), lane-change assistant, collision avoidance systems and parking aids. Implementation of these systems can considerably reduce the number of road accidents and mitigate the consequences. However, the low integration level and high cost of the commercially available modules to date, hamper the mass volume integration and standardization of these systems. Thus, there are research efforts, supported by the EU [2], to develop low-cost driver assistance systems that could be suitable also for low-budget cars.

The cost reduction can be achieved by high level integration of the building blocks on a single chip or in a package, referred to as system on chip (SoC) and system in package (SiP), respectively. Silicon-based technologies as CMOS or SiGe offer high on-chip integration capability and competitive performance compared to the III-V semiconductors as e.g. gallium-arsenide [3], which have been dominating the discrete microwave components market.

The standard digital CMOS process is particularly attractive, as it enables the high-level integration of analog and digital blocks. Recent advances in CMOS technology have enabled it to become an inexpensive alternative for realization of high-frequency integrated circuits. However, the required output power and temperature robustness, particularly for automotive radar sensor products, have not yet been achieved in standard digital CMOS technologies. Furthermore, metal-oxide-semiconductor (MOS) transistors suffer from very high flicker noise corner frequencies compared to bipolar transistors, making it difficult to build a direct down-

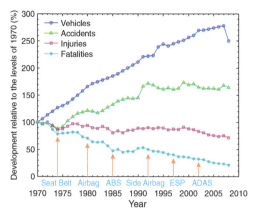

Fig. 1.1 Road traffic statistics in Germany.

conversion receiver in CMOS. This principle has the advantage of simplicity compared to the double-conversion or "sliding-IF" topologies.

The SiGe bipolar process offers transistors with excellent high-frequency characteristics, sufficient output power for automotive radar applications and the required robustness, but has the disadvantage of lower integration capability compared to CMOS. The use of a BiCMOS instead of a pure bipolar process resolves the integration drawback, but increases the costs and complexity.

The aim of this work is the realization of integrated receiver front-ends for narrow-band radar sensors at 24 GHz in Infineon's CMOS and SiGe technologies. Both technologies are automotive-certified and offer moderate mask costs at the present market volumes. These sensors can be useful for multiple car safety features such as lane-change assistant, side-crash detection, rear-collision warning or Stop and Go assistant, as presented in Fig. 1.2. Presently, some of the features are implemented using various approaches, such as CCD or CMOS cameras, ultrasonic sensors or lidar. Highly-integrated low-cost radar sensors may offer cheaper alternative for these features. Furthermore, realization of cost-effective sensors can enable their implementation in consumer and industrial applications as e.g. door openers, motion sensors and alarms.

Currently the market is dominated by ultra-wideband (UWB) 24 GHz short-range radar (SRR) sensors [4], [5]. However, according to Electronic Communications Committee's (ECC) decision, these sensors are allowed on the market in the EU only until July 2013 [6]. The allocated frequency range 21.625 – 26.625 GHz is only a temporary solution, whilst 79 GHz is intended for future SRR applications. However, the unlicensed industrial, scientific and medical (ISM) frequency range 24 – 24.25 GHz is an attractive alternative for mid-range radar sensors due to a higher allowable transmit power. Furthermore, it is still possible to use standard inexpensive packaging solutions [7], classical mounting techniques and moderately-priced measurement equipment at this frequency range. A frequency-modulated

1 Introduction

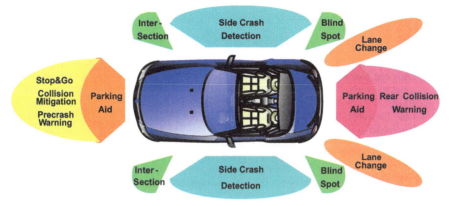

Fig. 1.2 Radar-based automotive safety features.

continuous-wave (FMCW) radar in the 24 GHz ISM band offering a 70 m outreach has been reported in [8].

Numerous publications report 24 GHz receivers in CMOS [9], [10] or SiGe [11], [12] technology. However, there are only a few publications that present fully ESD-protected receiver front-ends [13]. Sufficient ESD robustness and performance stability over a wide range of temperatures are required for hostile environment such as in automotive applications. Furthermore, there are few publications in the literature that offer direct comparison of receiver building blocks realized in different technologies [14].

This work presents the design, implementation, and characterization of building blocks and integrated receivers for 24 GHz narrow-band radar applications realized in CMOS and SiGe technologies. The performance stability of the circuits is compared over the extended automotive temperature range from -40 to 125 °C. The challenges posed to circuit design due to high ESD robustness requirements and corresponding circuit techniques are addressed. Furthermore, innovative circuit topologies for LNA, mixer and transceiver integration are proposed. Additionally, novel modeling and measurement techniques are presented.

This manuscript is organized as follows: chapter 2 provides an overview of radar principles, system architectures and the design challenges. The circuits in this work are realized in Infineon's CMOS and SiGe:C technologies, which are described in chapter 3. Modeling and simulation techniques are presented in chapter 4. Chapter 5 presents measurement techniques and discusses challenges of on-board measurement of differential devices. Circuit design and the experimental results of the building blocks and of the integrated receivers are described in chapter 6. Transceiver considerations and implementations are presented in chapter 7. Finally, chapter 8 summarizes the results and concludes this work.

4   1 Introduction

References

1. European Comission EC, "COMMUNICATION FROM THE COMMISSION - European Road Safety Action Programme - Halving the number of road accident victims in the European Union by 2010: A shared responsibility", http://eur-lex.europa.eu/LexUriServ/LexUriServ.do?uri=COM:2003:0311:FIN:EN:PDF, June 2003.
2. BMBF, "Projekt: Methoden zur zuverlässigen Systemintegration hochkompakter und kostenoptimaler 24 GHz Radarsensoren für Kfz-Anwendungen im Fahrerassistenzbereich - (MZS) - FAS-MZS", http://www.mstonline.de/foerderung/projektliste/detail_html?vb_nr=V3PID008, October 2006.
3. S. Voinigescu, D. S. McPherson, F. Pera, S. Szilagzyi, M. Tazlauanu, and H. Tran, "Comparison of Silicon and III-V Technology Performance and Building Block Implementations for 10 and 40 Gb/s Optical Networking ICs", *Journal of High Speed Electronics and Systems*, vol. 13, pp. 25--57, March 2003.
4. I. Gresham, A. Jenkins, R. Egri, C. Eswarappa, N. Kinayman, N. Jain, R. Anderson, F. Kolak, R. Wohlert, S. P. Bawell, J. Bennett, and J.-P. Lanteri, "Ultra-Wideband Radar Sensors for Short-Range Vehicular Applications", *IEEE Transactions on Microwave Theory and Techniques*, vol. 52, pp. 2105--2122, September 2004.
5. K.M. Strohm, H.-L. Bloecher, R. Schneider, and J. Wenger, "Development of future short range radar technology", *in European Radar Conference (EuRAD)*, pp. 165--168, Paris, France, October 2005.
6. Electronic Communications Committee ECC, "ECC Decision of 12 November 2004 on the frequency bands to be designated for the temporary introduction of Automotive Short Range Radars, ECC/DEC/(04)10", http://www.erodocdb.dk/Docs/doc98/official/pdf/ECCDEC0410.PDF, November 2004.
7. M. Engl, K. Pressel, J. Dangelmaier, H. Theuss, B. Eisener, W. Eurskens, H. Knapp, and W. Simbürger, "A 29 GHz Frequency Divider in a Miniaturized Leadless Flip-Chip Plastic Package", *in IEEE MTT-S International Microwave Symposium (IMS) Digest*, pp. 477--480, Fort Worth, USA, June 2004.
8. T. Wixforth and W. Ritschel, "Multimode-Radar-Technologie für 24 GHz", available at http://www.konstruktion.de/ai/resources/ed2055a235b.pdf, *Automobil Elektronik*, vol. 2, pp. 56-58, June 2004.
9. X. Guan and A. Hajimiri, "A 24-GHz CMOS front-end", *IEEE Journal of Solid-State Circuits*, vol. 39, pp. 368--373, February 2004.
10. Y.-H. Chen, H.-H. Hsieh, and L.-H. Hsieh, "A 24-GHz Receiver Frontend With an LO Signal Generator in 0.18-μm CMOS", *IEEE Transactions on Microwave Theory and Techniques*, vol. 56, pp. 1043--1051, May 2008.
11. S. Pruvost, L. Moquillon, E. Imbs, M. Marchetti, and P. Garcia, "Low Noise Low Cost Rx Solutions for Pulsed 24 GHz Automotive Radar Sensors", *in IEEE Radio Frequency Integrated Circuits (RFIC) Symposium Digest*, pp. 387--390, Honolulu, Hawaii, June 2007.
12. H. Veenstra, E. van der Heijden, M. Notten, and G. Dolmans, "A SiGe-BiCMOS UWB Receiver for 24 GHz Short-Range Automotive Radar Applications", *in IEEE MTT-S International Microwave Symposium (IMS) Digest*, pp. 1791--1794, Honolulu, Hawaii, June 2007.
13. S.-Y. Kim, K. V. Buer, E. Imbs, and G. M. Rebeiz, "An 18-20 GHz Subharmonic Satellite Down-Converter in 0.18 μm SiGe Technology", *in Topical Meeting on Silicon Monolithic Integrated Circuits in RF Systems (SiRF)*, pp. 1--4, San Diego, USA, January 2009.
14. X. Li, T. Brogan, M. Esposito, B. Myers, and K. K. O, "A comparison of CMOS and SiGe LNA's and mixers for wireless LAN application", *in Custom Integrated Circuits Conference (CICC)*, pp. 531--534, San Diego, USA, May 2001.

Chapter 2
Radar Systems

Automotive safety systems require information about the objects in the vicinity of the vehicle. These data are usually obtained by sensing the surroundings. A typical sensor system usually transmits a signal and estimates the attributes of the available targets, such as velocity or distance from the sensor, based on the measurement of the scattered signal. The signal used for this purpose in radar (radio detection and ranging) systems is an electromagnetic (EM) wave at microwave frequencies. The main advantage of radar systems compared to other alternatives such as sonar or lidar is the immunity to weather conditions and potential for lower cost realization.

Section 2.1 describes the principle of radar. There are two main operation principles, continuous wave (CW) and pulsed. The latter is not treated within this scope, since the frequency regulations in the ISM band result in a limitation on the absolute transmitter power. Thus, pulsed radar would result in a lower SNR due to a lower duty cycle compared to the CW operation. Radar operation is discussed in sections 2.2 - 2.4. Frequency regulations around 24 GHz are described in section 2.5. Typical radar architectures and circuit related challenges are presented in section 2.6. Section 2.7 provides an overview of the automotive radar systems and their application for car safety. Finally, section 2.8 concludes this chapter with considerations on technology features needed for radar realization.

2.1 Radar Principle

Radar systems are composed of a transmitter that radiates electromagnetic waves of a particular waveform and a receiver that detects the echo returned from the target. Only a small portion of the transmitted energy is re-radiated back to the radar, which is then amplified, down-converted and processed. The range to the target is evaluated from the travelling time of the wave. The direction of the target is determined by the arrival angle of the echoed wave. The relative velocity of the target is determined from the doppler shift of the returned signal.

V. Issakov, *Microwave Circuits for 24 GHz Automotive Radar in Silicon-based Technologies*, DOI 10.1007/978-3-642-13598-9_2, © Springer-Verlag Berlin Heidelberg 2010

For automotive radar applications the separation between the transmitter and receiver is negligible compared to the distance to a target. Thus, these systems are monostatic in a classical sense. However, the automotive radar systems are usually referred to as *bistatic* when two separate antennas are used for transmit and receive and *monostatic* when the same antenna is used for these functions, as depicted in Fig. 2.1. The latter configuration requires a duplexer component to provide isolation between transmitter and receiver. This is usually realized using expensive external bulky transmit/receive (T/R) switch or circulator components. The solution of using hybrid ring coupler [1] offers a cost advantage at the expense of lower performance due to higher losses and increased noise figure.

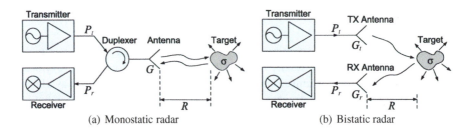

Fig. 2.1 Radar configurations.

2.2 Radar Equation and System Considerations

The radar equation provides the received power level as function of the characteristics of the system, the target and the environment. The well-known bistatic radar equation [2] for the system in Fig. 2.1(b) is given by

$$P_r = \frac{P_t A_{er} A_{et} \sigma}{4\pi R^4 \lambda^2 L_{sys}}, \tag{2.1}$$

where P_r is the received power, P_t is the transmitted power, A_{er} and A_{et} are the effective area of the receive and transmit antennas, respectively, R is the distance to the target, σ is the radar cross-section (RCS), defined as the ratio of the scattered power in a given direction to the incident power density and L_{sys} is the system loss due to misalignment, antenna pattern loss, polarization mismatch, atmospheric loss [3], but also due to analog to digital conversion and fast Fourier transform (FFT) windowing. Taking into consideration that the effective area of the receive and transmit antenna is related to the wavelength λ and to the antenna gain G_r and G_t, as $A_{er} = G_r \lambda^2/4\pi$ and $A_{et} = G_t \lambda^2/4\pi$, respectively, the radar equation can be rewritten as

$$P_r = \frac{P_t G_r G_t \lambda^2 \sigma}{(4\pi)^3 R^4 L_{\text{sys}}}. \tag{2.2}$$

Based on the system characteristics and the noise floor of the receiver a certain minimal signal power level $P_{\text{r,min}}$ is required in order to detect the target. Thus, from (2.2) the maximum achievable radar range can be calculated as follows

$$R_{\text{max}} = \left(\frac{P_t G_r G_t \lambda^2 \sigma}{(4\pi)^3 P_{\text{r,min}} L_{\text{sys}}} \right)^{1/4}. \tag{2.3}$$

Furthermore, in most practical designs a minimal signal to noise ratio (SNR) at the output of the receiver $\text{SNR}_{\text{o,min}}$ is considered in order to ensure high probability of detection and low false-alarm rate. Typically, SNR values of higher than 12 dB are required. The noise factor of a receiver is defined as

$$F = \frac{S_i/N_i}{S_o/N_o}, \tag{2.4}$$

where S_i and S_o are the input and output signal levels, respectively, N_o is the noise level at the receiver output and N_i is the input noise level, given by

$$N_i = k_B T B, \tag{2.5}$$

where B is the system bandwidth, k_B is the Boltzmann constant and T is the temperature in Kelvin. Taking into consideration that there is an additional processing gain due to the integration over several pulses, approximately given by $G_{\text{int}} = T_{\text{CPI}} \cdot B$, where T_{CPI} is the coherent processing interval (CPI), the maximum radar range in (2.3) can be rewritten as a function of $\text{SNR}_{\text{o,min}}$ as follows

$$R_{\text{max}} = \left(\frac{P_t G_r G_t \lambda^2 \sigma T_{\text{CPI}}}{(4\pi)^3 \cdot kTF \cdot \text{SNR}_{\text{o,min}} \cdot L_{\text{sys}}} \right)^{1/4}. \tag{2.6}$$

The attenuation for the propagation of the electromagnetic waves at 24 GHz is about 0.15 dB/km [4]. Taking into consideration that the typical range for automotive radar sensors is up to 200 m, the contribution of the atmospheric attenuation to L_{sys} is negligible. Even under heavy rain or fog conditions the attenuation over these distances is in the range of few decibels.

The RCS of typical targets in automotive applications ranges from 0.1 to 200 m^2. The antenna gain is usually in the range of 15 - 25 dBi. Antennas are typically realized as patch antenna arrays for beam shaping. Their large size at 24 GHz limits the dimensions of radar modules.

Equation (2.6) can be rearranged for the noise factor F. Plugging in the smallest RCS and the largest required distance of operation results in the required receiver noise figure. For example, for an object with a σ of 0.1 that has to be detected at a maximal distance of 100 m with transmit and receive antenna gains of 20 dB, transmitter power of 0 dBm, the system losses of 3 dB, the CPI time of 2 ms and minimum required SNR after the FFT of 12 dB, the required receiver front-end noise figure is 10.75 dB. For a typical narrow-band 24 GHz system a single side-band (SSB) noise figure (NF) of less than 10 dB is needed. The NF is related to the noise factor in (2.4) as $\text{NF} = 10 \cdot \log(F)$. The gain of a receiver front-end is less crucial, since it can be compensated in the baseband stage. However, it still has to be above 10 dB for a low receiver NF, due to noise figure cascading.

Another limiting case, referred to as the *blocker* case, is the scenario of a large target with maximum RCS being present very close to a radar at a minimal distance of operation. This sets the requirement on the front-end linearity in terms of input-referred 1dB compression point (IP1dB), which should be typically above -15 dBm. Combination of both mentioned limiting cases results in a requirement on the receiver's dynamic range (DR), which usually should be above 70 dB.

2.3 CW and Frequency-Modulated Radar

2.3.1 Doppler Radar

A classical continuous wave (CW) or Doppler radar implementation uses a fixed transmit frequency to detect a moving target and its velocity. It is based on the Doppler frequency shift. If there is a non-zero relative velocity v_r between a radar transmitter sending a signal at frequency f_0, and a moving target, the returned signal has frequency $f_0 + f_d$, where f_d is the Doppler frequency shift given by

$$f_d = \frac{2v_r}{c} f_0, \tag{2.7}$$

where c is the speed of light. The relative velocity v_r of a target is determined by the velocity component along the line-of-sight of the radar and is given by

$$v_r = v_a \cos \theta, \tag{2.8}$$

where v_a is the actual velocity of a target and θ is the angle between the target trajectory and the line-of-sight, as depicted in Fig. 2.2.

It can be observed from (2.8) that for an acute angle $\theta < 90°$, corresponding to an approaching target, the Doppler shift is positive $f_d > 0$ and for an obtuse angle $\theta > 90°$, corresponding to a receding target, the Doppler shift is negative $f_d < 0$. Furthermore, for $\theta = 90°$ the Doppler shift is zero. Thus, the velocity component perpendicular to the line-of-sight cannot be determined.

2.3 CW and Frequency-Modulated Radar

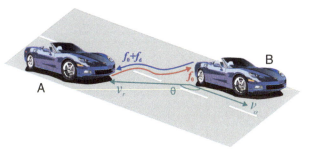

Fig. 2.2 Doppler effect.

2.3.2 Frequency-Modulated Radar

A simple CW radar allows determination of target velocity, but not the distance to target. Therefore, frequency-modulated continuous-wave (FMCW) systems have been developed to resolve this drawback. There are several possibilities to modulate a carrier frequency in time, such as linear frequency modulation (LFM), frequency shift-keying (FSK) or frequency-stepped continuous-wave (FSCW) modulation. This section describes the commonly implemented LFM and FSCW modulation schemes.

2.3.2.1 Linear FM Continuous-Wave Radar

The most common modulation technique is LFM that modulates the transmit frequency with a triangular waveform [5]. The principle is exemplified in Fig. 2.3 showing the waveforms for a target at a distance R, approaching with a relative velocity v_r.

The transmit signal varies between the minimum frequency f_0 and the maximum frequency $f_0 + B$ with a period T_m, where B is the bandwidth. At time t_1, the transmitter sends a signal with frequency f_1. This signal is received at time t_2, after a round-trip delay of $\tau = 2R/c$, with a frequency shifted by f_d. Meanwhile, the transmitter frequency is f_2. A mixer produces a base band signal at the instantaneous difference frequency between the transmit f_{TX} and the receive f_{RX} signals, referred to as the beat signal $f_b = |f_{TX} - f_{RX}|$. When the target is stationary, the beat frequency level is only related to the range and is given by

$$f_R = f_{b,\text{stationary}} = \frac{4B}{T_m} \cdot \frac{R}{c}. \tag{2.9}$$

For a moving target the Doppler effect shifts the absolute values of the received frequencies. Thus, the down-converted frequencies are

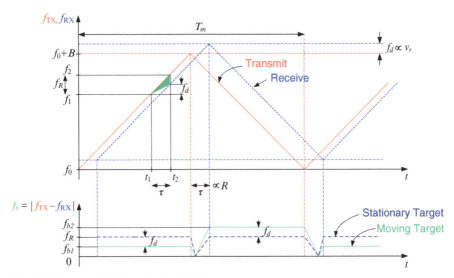

Fig. 2.3 Range and relative velocity detection.

$$f_{b1} = f_R - f_d, \qquad (2.10)$$
$$f_{b2} = f_R + f_d, \qquad (2.11)$$

for the rising and for the falling slope of the transmit signal, respectively. Once the beat frequency levels f_{b1}, f_{b2} have been measured in the baseband, the range and the relative speed can be calculated by [2]

$$R = \frac{c(f_{b1} + f_{b2})T_m}{8B}, \qquad (2.12)$$

$$v_r = \frac{c(f_{b2} - f_{b1})}{4f_0}, \qquad (2.13)$$

where f_0 is the frequency of the transmitted signal.

Since the beat frequency signal is a rectangular signal with a period $T_m/2$, the corresponding spectrum is a sinc function centered at f_b and the first zero crossing occurs at $2/T_m$. Thus, the smallest resolvable frequency Δf is the reciprocal of the measurement time

$$\Delta f = \frac{2}{T_m}. \qquad (2.14)$$

Substituting (2.14) into (2.7) the minimal resolvable velocity is obtained

$$\Delta v_r = \frac{c}{2f_0} \cdot \Delta f = \frac{c}{f_0} \frac{1}{T_m}. \qquad (2.15)$$

2.4 Angle Detection

Thus, for larger T_m or lower modulation frequency, higher velocity resolution can be achieved. Additionally, substituting (2.14) into (2.9) one obtains expression for the range resolution

$$\Delta R = \frac{cT_m}{4B} \cdot \Delta f = \frac{c}{2B}. \tag{2.16}$$

As can be observed, larger bandwidth offers higher range resolution. Thus, the range resolution of a narrow-band radar is limited.

Further insight on the FMCW radar using linear frequency modulation is presented in Appendix A. Additionally, a more advanced modulation algorithm FSCW that combines LFM and FSK techniques is briefly described in Appendix B.

2.4 Angle Detection

The monopulse principle can be described on two antennas having complex receive patterns $G_1(\alpha)$ and $G_2(\alpha)$. The distance between the antennas is d, as shown in Fig. 2.4. The phase difference $\Delta\varphi$ for an incident plane wave is

$$\Delta\varphi = d\sin(\alpha)\frac{2\pi}{\lambda}, \tag{2.17}$$

where $\lambda = c/f_0$ is the wavelength, related to the carrier frequency f_0.

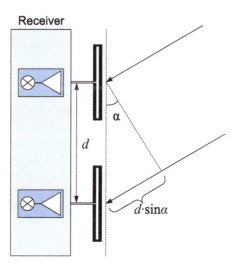

Fig. 2.4 Antenna array in receive mode.

The difference and sum of the received signals of both antennas are given by

$$\Delta(\alpha) = G_1(\alpha) - e^{-i\Delta\varphi} \cdot G_2(\alpha), \tag{2.18}$$

$$\Sigma(\alpha) = G_1(\alpha) + e^{-i\Delta\varphi} \cdot G_2(\alpha). \tag{2.19}$$

For the monopulse angle detection the ratio $R_{\text{mono}} = \Delta/\Sigma$ is considered.

If the detection is based only on the amplitude of R_{mono}, it is called *amplitude-comparison monopulse*. This approach uses two overlapping antenna beams, so that the radiation patterns have slightly different look directions. This technique is preferred for long-range radars (LRR) with a small coverage angle, as e.g. for the 24 GHz LRR in [6].

If the antenna patterns G_1 and G_2 are identical, only the phase difference can be used for angle detection. This is referred to as the *phase-comparison monopulse* and or as the *phase interferometry*. In this case the ratio R_{mono} becomes

$$R_{\text{mono}} = \frac{\Delta}{\Sigma} = \frac{1 - e^{i\Delta\varphi}}{1 + e^{i\Delta\varphi}}. \tag{2.20}$$

In the phase monopulse technique the phase difference $\Delta\varphi$ is evaluated in order to avoid the amplitude calibration, required in the amplitude monopulse technique. The angle of arrival is then easily obtained by rearranging equation (2.17)

$$\alpha = \sin^{-1}\left(\frac{\lambda\Delta\varphi}{2\pi d}\right). \tag{2.21}$$

The phase monopulse technique is preferred for 24 GHz systems, because the antennas are implemented as patch antennas orientated in the same look direction, as e.g. in [7].

The unambiguous angular range depends on the distance d between two receive elements

$$\Delta\alpha = 2 \cdot \sin^{-1}\left(\frac{\lambda}{2 \cdot d}\right). \tag{2.22}$$

For short-range applications the unambiguous angular range should be very close to $+/-90°$. Therefore, the spatial sampling theorem has to be fulfilled and the antenna separation has to be half-wavelength. For mid-range and long-range systems the spacing has to be chosen according to the beamwidth of the transmitter antenna. An increased spacing between the receiver antennas allows to increase the size of the antennas and thus the gain. Furthermore, this results in a direct improvement of the angle measurement accuracy. However, this results in an increased radar module size.

2.5 Frequency Regulations

The performance of radar systems and the applied waveform principles are strongly influenced by the frequency regulations. The maximum allowable power limits and

2.5 Frequency Regulations

the corresponding measurement procedures for 24 GHz radar systems are defined in the ETSI standard EN 302 288-1 [8]. This document defines the spectral mask of the maximum allowed transmitter power in the ISM and UWB frequency bands around 24 GHz, as shown in Fig. 2.5.

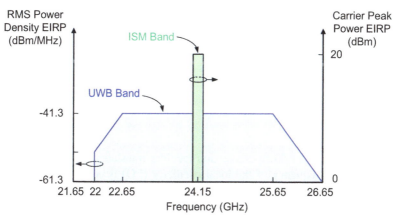

Fig. 2.5 Transmitter maximum radiated power spectral mask.

The limit for the transmitted power is given as equivalent isotropic radiated power (EIRP). The EIRP value is given in dBm by adding the gain of the transmitter antenna to the actual transmitter power

$$P_{\text{EIRP}}(\text{dBm}) = P_{\text{TX}}(\text{dBm}) + G_{\text{TX}}(\text{dB}). \qquad (2.23)$$

In the ISM band from 24.05 GHz to 24.25 GHz the maximum power is constrained to 20 dBm. For the ultra-wide band from 22.65 GHz to 25.65 GHz a maximum power spectral density of only -41.3 dBm/MHz is allowed. This spectral density is very low and can only be used by pulsed systems with high bandwidth.

The ISM band of 200 MHz is applicable for automotive mid-range to long-range applications, since these are typically implemented as continuous wave (CW) systems and require higher power to achieve the necessary maximum range. CW systems provide a higher SNR compared to pulsed systems for the same transmitter power. Furthermore, the narrow available bandwidth of the ISM band is practical, since the maximum bandwidth is limited by the required signal to noise ratio at the maximum range [2]. Apart from the automotive radar systems, the ISM band is used for less demanding applications, such as door openers or surveillance. These systems are implemented using Doppler sensors without any additional frequency modulation.

2.6 Receiver Architectures

A receiver is used to amplify and down-convert a radio frequency (RF) signal with minimal added distortion. Therefore, the requirements for the receiver performance are usually very demanding. It should offer low noise, high dynamic range, and high local oscillator (LO) isolation so as to avoid radiation emission. The choice of receiver architecture is usually determined by complexity, power dissipation and system considerations. Receiver architectures can be classified with respect to the down-conversion topology.

2.6.1 Homodyne

A homodyne or direct down-conversion receiver translates an RF signal directly to zero-IF. The frequency of the local oscillator (LO) is equal to the carrier frequency of the received RF signal. This architecture offers the advantage of simplicity. Avoiding an additional down-conversion to an intermediate frequency (IF) saves chip area, current consumption, complexity and avoids the image rejection problem inherent to heterodyne systems. A simplistic block diagram of a continuous-wave radar system implementing a homodyne receiver is presented in Fig. 2.6. The spectrum diagram of a homodyne receiver is depicted in Fig. 2.7. Both upper and lower side bands are down-converted to zero-IF.

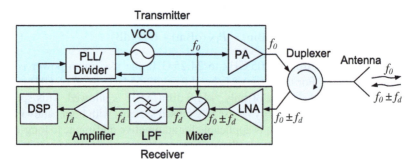

Fig. 2.6 Block diagram of a homodyne CW radar system.

However, it has a serious disadvantage when implemented in CMOS technology. The expected baseband frequencies for a 24 GHz FMCW radar are in the range from 1 kHz to 100 kHz, whilst the flicker noise corner frequency of the CMOS circuits is around 10 MHz. Thus, for an active mixer implementation very high noise figures of above 40 dB at 1 kHz can occur. Therefore, either implementation of advanced circuit techniques [9] or of passive mixers [10] is required to resolve this issue. This problem will be addressed in detail in chapter 6. Bipolar transistors have a

2.6 Receiver Architectures

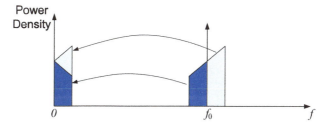

Fig. 2.7 Frequency translation of a homodyne receiver.

much lower flicker noise corner frequency and therefore may be suitable for zero-IF receivers. However, there is an additional disadvantage. Parasitic DC signals appear due to mismatch, LO self-mixing and RF crosstalk [11]. The DC offset can be suppressed by DC blocking capacitors. The high-pass characteristics offered by these capacitors also act as a sensitivity time control (STC), which suppresses low-frequencies generated by nearby targets.

2.6.2 Heterodyne

A heterodyne receiver down-converts an RF signal to an intermediate frequency, which is typically in the range of few gigahertz. Implementation of the IF frequency mitigates the flicker noise problem and allows for better selectivity due to an easier bandpass filter realization at the IF. However, it requires more circuit blocks and an additional IF reference frequency. A conceptual simplified block diagram of a CW radar using heterodyne architecture is presented in Fig. 2.8. The spectrum diagram of a heterodyne receiver is depicted in Fig. 2.9.

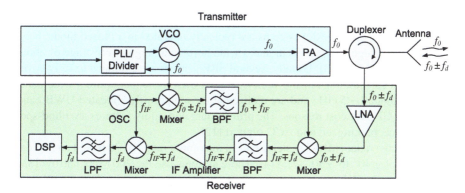

Fig. 2.8 Block diagram of a heterodyne CW radar system [2].

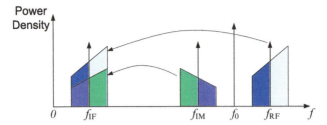

Fig. 2.9 Frequency translation of a heterodyne receiver.

As can be seen, both the wanted and the image bands are down-converted to the IF frequency. Thus, the image frequency has to be suppressed before it is mixed down to the IF. This requires a bandpass filter. For a practical filter quality factor, the IF frequency should be sufficiently high, so that the RF is far from the LO frequency.

If the IF is low, image suppression at the RF becomes impossible, but signals could be processed directly at low frequencies of few megahertz and the image suppression performed at the IF [11]. For automotive radar systems in SiGe a direct down-conversion is the preferred option due to lower power consumption and complexity, whilst for CMOS a low-IF solution is preferred.

2.7 Status of Automotive Radar Systems

There are various ways of grouping the commercially available radar systems for automotive applications, e.g. by their bandwidth (narrow-band or wide-band), by their operation principle (pulsed or continuous wave) or by the covered area (short-range, mid-range and long-range radar). This section presents a brief overview of the current automotive radar implementations and classifies them with respect to their operating range and typical applications.

Short-range radar (SRR) systems are typically operated in a pulsed mode, have a maximum range of up to 30 m and a wide horizontal angular coverage of about $\pm 65°$ to $\pm 80°$ [12], [13]. Usually, several SRR sensors are equipped to fully cover the nearest surroundings of the vehicle. Pulsed SRR systems require a wide bandwidth of about 3 – 5 GHz and are realized at the temporarily allocated UWB band around 24 GHz for cost reasons. The primary targeted safety features are blind-spot surveillance, parking aid and ACC support [12].

Mid-range radar (MRR) systems have a maximum range of 70 m and an angular coverage of $\pm 40°$ to $\pm 50°$ [7], [14]. These systems use the 200 MHz narrow ISM band around 24 GHz and operate in continuous wave mode using linear frequency modulation (LFM) or advanced modulation techniques such as e.g. frequency shift keying (FSK) or frequency-stepped continuous wave (FSCW) [15]. Due to the low available bandwidth the range resolution is limited to 0.6 m. Therefore, the primary targeted application for these sensors is the lane-change assistant.

Long-range radar (LRR) systems have a maximum range of up to 200 m and an angular coverage of $\pm 4°$ to $\pm 8°$ [16]. Most of the commercially available systems use the allocated $76 - 77$ GHz frequency range and operate in continuous wave mode using FMCW. However, there are also systems available that offer similar functionality using a narrow-band 24 GHz radar [14]. The long-range sensors are implemented typically for ACC.

2.8 Technology Requirements for Radar Chipset

Based on the radar system considerations presented in section 2.1, the integrated circuits for a radar front-end should fulfill demanding performance requirements around the frequency of 24 GHz. This poses a challenge on circuit design, which can be relaxed by implementation of high performance technologies, based on III-V semiconductor compounds such as e.g. gallium-arsenide (GaAs) or indium-phosphide (InP) [17]. These technologies can provide transistors with very high output power, very low noise, high gain and good linearity. However, they have a low integration level, which results in increased bill of materials (BOM) and module assembly costs. Implementation of circuits in cheaper silicon-based technologies offers the advantage of high integration, but at the cost of lower performance. Thus, in order to achieve sufficient performance, implementation of advanced circuit techniques is required.

Until now the commercial radar sensors have used front-end chip sets realized mainly in the III-V semiconductor technologies [18], [19]. The implementations usually comprise many discrete components. Therefore, the new generation of radar sensors uses integrated circuits based on SiGe technology [20]. Future generations of radar circuit implementation would further take advantage of high integration capabilities of CMOS. Transceiver front-ends for automotive applications have been already demonstrated at 24 GHz and at 77 GHz in BiCMOS technology [21].

References

1. B. Dehlink, *Integrated Millimeter Wave Front-End Design in SiGe Bipolar Technology*, Dissertation, Institut für Nachrichten- und Hochfrequenztechnik der TU Wien, 2007.
2. K. Chang, *RF and Microwave Wireless Systems*, Wiley, 2000.
3. M. Skolnik, *Introduction to Radar Systems*, McGraw-Hill, 1981.
4. Naval Air Warfare Center US Navy, *Electronic Warfare and Radar Systems Engineering Handbook*, http://www.microwaves101.com/encyclopedia/Navy_Handbook.cfm, 1999.
5. A. G. Stove, ''Linear FMCW radar techniques'', *IEE Proceedings F, Radar and Signal Processing*, vol. 139, pp. 343--350, October 1992.

6. V. Cojocaru, H. Kurata, D. Humphrey, B. Clarke, T. Yokoyama, V. Napijalo, T. Young, and T. Adachi, "A 24 GHz Low-Cost, Long-Range, Narrow-Band, Monopulse Radar Front End System for Automotive ACC Applications", *in IEEE MTT-S International Microwave Symposium (IMS) Digest*, pp. 1327--1330, Honolulu, USA, June 2007.
7. R. Mende, "UMRR: A 24 GHz Medium Range Radar Platform", http://smartmicro. de/UMRR_-_A_Medium_Radar_Radar_Platform.pdf, July 2003.
8. European Telecommunications Standards Institute ETSI, "European Standard EN 302 288-1 Electromagnetic Compatibility and Radio Spectrum Matters (ERM); Short Range Devices; Road Transport and Traffic Telematics (RTTT); Short Range Radar Equipment Operating in the 24 ghz Range; Part 1: Technical Requirements and Methods of Measurement", http://www.etsi.org/WebSite/Technologies/AutomotiveRadar. aspx, May 2006.
9. H. Darabi and J. Chiu, "A Noise Cancellation Technique in Active RF-CMOS Mixers", *IEEE Journal of Solid-State Circuits*, vol. 40, pp. 2628--2632, Dec 2005.
10. R. M. Kodkani and L. E. Larson, "A 24-GHz CMOS Passive Subharmonic Mixer/Downconverter for Zero-IF Applications", *IEEE Transactions on Microwave Theory and Techniques*, vol. 56, pp. 1247--1256, May 2008.
11. J. Crols and M. Steyaert, *CMOS Wireless Transceiver Design*, Springer, 1997.
12. K.M. Strohm, H.-L. Bloecher, R. Schneider, and J. Wenger, "Development of future short range radar technology", *in European Radar Conference (EuRAD)*, pp. 165--168, Paris, France, October 2005.
13. J. Wenger, "Short range radar - being on the market", *in European Radar Conference (EuRAD)*, pp. 255--258, Munich, Germany, October 2007.
14. R. Weber and N. Kost, "24-GHz-Radarsensoren für Fahrerassistenzsysteme", *ATZ Elektronik*, vol. 2, pp. 16--22, 2006, http://www.atzonline.de/Artikel/3/3349/ 24-GHz-Radarsensoren-fuer-Fahrerassistenzsysteme.html.
15. H. Rohling and M.-M. Meinecke, "Waveform design principles for automotive radar systems", *in CIE International Conference on Radar*, pp. 1--4, Beijing, China, October 2001.
16. M. Schneider, "Automotive Radar Status and Trends", *in German Microwave Conference (GeMiC)*, pp. 144--147, Ulm, Germany, April 2005.
17. J. Godin, M. Riet, S. Blayac, P. Berdaguer, J.-L. Benchimol, A. Konczykowska, A. Kasbari, P. Andre, and N. Kauffman, "Improved InGaAs/InP DHBT Technology for 40 Gbit/s Optical Communication Circuits", *in IEEE GaAs IC Symposium Technical Digest*, pp. 77--80, Seattle, USA, November 2000.
18. J.-E. Müller, T. Grave, H. J. Siweris, M. Kärner, A. Schäfer, H. Tischer, H. Riechert, L. Schleicher, L. Verweyen, A. Bangert, W. Kellner, and T. Meier, "A GaAs HEMT MMIC Chip Set for Automotive Radar Systems Fabricated by Optical Stepper Lithography", *IEEE Journal of Solid-State Circuits*, vol. 32, pp. 1342--1349, September 1997.
19. R. Troppmann and A. Höger, "ACC-Systeme Hardware, Software und Co. - Teil 2", available at www.hanser-automotive.de/fileadmin/heftarchiv/2004/ 4918.pdf, *Hanser Automotive*, vol. 3, pp. 58-62, May 2005.
20. W. Lehbrink, "Radar-Chips aus SiGe", available at www.hanser-automotive.de/ uploads/media/24380.pdf, *Hanser Automotive*, vol. 2, pp. 14-18, March 2008.
21. V. Jain, F. Tzeng, L. Zhou, and P. Heydari, "A Single-Chip Dual-Band 22-to-29 GHz/77-to-81 GHz BiCMOS Transceiver for Automotive Radar", *in IEEE International Solid-State Circuits Conference (ISSCC)*, pp. 308--309, San Francisco, February 2009. IEEE.

Chapter 3
CMOS and Bipolar Technologies

Performance capability of circuits depends to a great extent on the available semiconductor technology. Numerous parameters related to device, metallization or substrate properties circumscribe the achievable circuit characteristics at microwave frequencies and thus can be used for technology comparison.

The circuits presented in this work have been designed and processed in Infineon's standard 0.13 µm digital CMOS C11N and 0.35 µm bipolar SiGe:C B7HF200 technologies. The C11N process has been chosen due to the cost advantage compared to advanced CMOS technology nodes, whilst B7HF200 technology is used due to excellent high-frequency performance and high robustness. Both technologies are automotive qualified and suitable for realization of radar circuits.

Sections 3.1 and 3.2 describe the implemented technologies and provide a brief overview of the available devices. Section 3.3 concludes this chapter with a direct comparison of device performance and process key parameters.

3.1 CMOS Technology

Fig. 3.1 describes a modern dual-well sub-micrometer CMOS technology. It offers

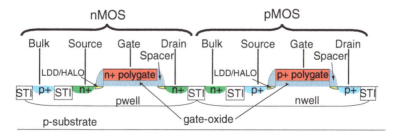

Fig. 3.1 Cross-section of a sub-micrometer CMOS process [1].

two basic types of transistors - the *n*-channel and the *p*-channel MOSFET. The structure in Fig. 3.1 is similar to a typical front end of line (FEOL) of Infineon's 0.13 μm C11N technology [2].

The *p*-well and *n*-well are formed in a *p*-type substrate without an additional intermediate epitaxial layer, since the C11N process is realized on a non-*epi* substrate [3]. The NMOS and PMOS transistors are realized in the low-doped *p*-well and *n*-well, respectively. The well can be contacted and is usually referred to as the *bulk* or the *body*. The *source* and *drain* terminals of the devices are realized as heavily doped n-regions for NMOSFET and *p*-regions for PMOSFET, respectively. A heavily doped conductive polysilicon acts as the *gate* terminal. Silicide is typically used to form ohmic contacts for source, drain and gate. A thin silicon-dioxide (SiO_2) layer, also referred to as the *gate oxide*, insulates the gate from the substrate. Shallow trench isolation (STI), realized using SiO_2, is implemented in order to reduce leakage between transistors and thus allow higher transistor density.

As can be seen in Fig. 3.1, for an NMOS transistor both the well and the substrate are *p*-doped. Thus, the bulk contacts of all NMOS transistors on one substrate are tied to the same electric potential. Therefore, in a dual-well CMOS process it is not possible to control the bulk potential of each NMOS transistor separately, as in the case of PMOS transistors. In this work the bulks of NMOS transistors are tied to the ground potential, whilst the bulks of PMOS transistor are connected to the source in order to reduce bulk effects.

As the device dimensions continue to scale down in the modern CMOS submicrometer technologies, numerous short-channel effects, such as drain-induced barrier lowering (DIBL) or hot-electron effects [4] degrade the transistor performance and reliability. Therefore, the C11N technology also implements a lightly doped drain (LDD) implant in order to reduce the peak electric field and mitigate the hot-electron effects. Furthermore, HALO doping is used in order to increase the average channel doping concentration and thus to increase the threshold voltage and compensate for the DIBL effect.

3.1.1 MOSFET Layout and Modeling Considerations

The classical basic MOSFET equations [5], are widely used for manual circuit calculations. However, 0.13 μm transistors are well in the sub-micrometer region and thus exhibit short-channel effects [6]. As a result, this simplistic model provides only a very rough approximation. Additionally, solving a circuit with several transistors by means of an equivalent model including all parasitics is a very cumbersome matter. Thus, performance evaluation at the design stage can be performed only in simulation using a considerably more complex transistor model, such as e.g. an EKV, a higher level HSPICE or a BSIM model. The circuits in this work have been simulated using enhanced BSIM 4.0 model [7].

Careful modeling of transistor parasitics is essential for accurate simulation of microwave circuits. This is particularly important for narrow-band designs, since

3.1 CMOS Technology

the circuits have to achieve the center frequency very accurately for optimum performance. A BSIM model includes only the internal transistor parasitics. Thus, it has to be extended by layout-dependent external parasitics, as shown in Fig. 3.2.

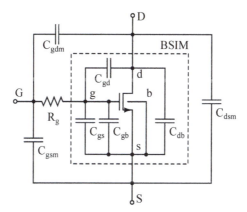

Fig. 3.2 Extended MOSFET simulation model.

The parasitic capacitances C_{gsm}, C_{gdm} and C_{dsm} describe the capacitances between the metal interconnects leading to the transistor terminals. The physical dimensions of the interconnects and thus the capacitance values depend on the transistor size, amount of layers used for routing, maximum allowed current density of the metal, maximum output power from the transistor and minimum allowed spacing. Values of these capacitances can reach those of intrinsic capacitances. The gate resistance R_g depends on the layout topology and on gate contacts.

CMOS transistors for high-frequency designs are usually implemented using a multifinger layout, as shown in Fig. 3.3. This folded structure is advantageous for

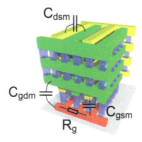

Fig. 3.3 Multifinger transistor layout.

a compact layout, whilst the same W/L ratio is maintained. The external parasitic elements are annotated on the drawing. As may be inferred from the figure, using more routing layers results in an increased metallization capacitance. Furthermore, the capacitance increases for dense routing using minimal separation between the traces. Therefore, advanced asymmetrical routing techniques for multifinger layout have been proposed in [8].

The values of the external capacitances C_{gsm}, C_{gdm} and C_{gsm} are usually obtained numerically using an electromagnetic 3D field solver. The gate resistance R_g for the multifinger layout is calculated analytically by [9]

$$R_g = \frac{R_{sh}}{12n_f^2} \frac{W}{L} + \frac{\rho_{con}}{WL},$$
(3.1)

where R_{sh} is the sheet resistance of the gate material, typically about $7\,\Omega/\square$ for polysilicon, n_f is the number of fingers for multifinger layout of transistor, W and L are the total width and gate length of the transistor, respectively, and ρ_{con} is the contact resistivity of the interface between the silicide and the polysilicon. The factor 12 in the denominator of the first term is valid when the gate is contacted on both sides. When the gate is contacted on one side, this factor becomes 3.

The gate resistance contributes to thermal noise and degrades the high-frequency performance of a transistor. Thus, it should be minimized for the design of microwave circuits, particularly of an LNA. The only technology-independent parameters in (3.1) that can be optimized by layout are n_f, W and L. The gate length L is usually set to the minimum allowed by technology for highest transistor speed. The first term in (3.1) can be reduced by using a higher number of fingers. Additionally, increasing the width W reduces the second term, but increases the first term and parasitic capacitance. Thus, the layout can be optimized for lowest R_g.

3.1.2 Devices Available in C11N

Infineon's C11N process offers transistors with oxide thicknesses of 2.2 nm and 5.2 nm, having maximum operation voltages of 1.5 V and 3.3 V, respectively. Transistors with low, regular and high threshold voltage V_t are provided. Highest performance can be achieved at 1.5 V using 2.2 nm devices [2]. For stacked circuit topologies low-V_t transistors are preferred due to relaxed headroom.

Several types of polysilicon, diffusion, n-well and metal resistors are provided. The lowest parasitics and thus the best high-frequency performance are achieved using polysilicon resistors. The non-silicided p^+ polysilicon resistor offers a sheet resistance of $300\,\Omega/\square$ and an excellent temperature coefficient of $0.01\%/^\circ$C.

C11N provides several types of capacitors, such as MOS, diffusion and the metal-insulator-metal (MIM). The latter has very low parasitics and a high capacitance density of $1.042\,\mathrm{fF}/\mu\mathrm{m}^2$. Additionally, it offers high linearity, low losses, good matching and reliability. Therefore, it is used throughout this work.

3.2 Bipolar Transistors

The cross-section of an *npn* transistor in Infineon's B7HF200 SiGe:C technology is presented in Fig. 3.4. The process is similar to the one described in [10].

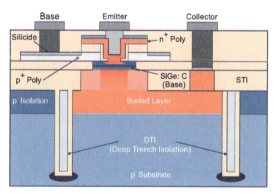

Fig. 3.4 Cross-section of an *npn* transistor [11].

It is based on a double-polysilicon self-aligned transistor concept [12] that offers low parasitic capacitances and a low extrinsic base resistance. The active area of the transistor in Fig. 3.4, located below the emitter contact, is small compared to transistor dimensions. The SiGe:C base layer is grown by a selective epitaxy. The base is contacted by a silicided p^+ polysilicon layer, which offers low external base resistance. The collector attached through a highly doped low-ohmic buried layer, which is then contacted by a highly doped n^+ region. The mono-crystalline emitter contact offers a small emitter resistance. Implementation of deep trench isolation (DTI) reduces collector-substrate C_{cs} and base-collector C_{bs} capacitances, whilst STI is used to minimize the transistor dimensions.

The drawn width of the emitter mask window in B7HF200 is 0.35 μm and results in relaxed lithography, whilst an effective emitter width of 0.18 μm is achieved using spacers between emitter and base contacts.

Heterojunction bipolar transistors (HBT) in B7HF200 implement a silicon germanium (SiGe) compound in the base in order to modify the band diagram compared to a classical bipolar homojunction transistor (BJT). The narrower band-gap of SiGe compared to Si results in a lower barrier for injection of emitter majority carriers (electrons) into the base, and a higher barrier for back injection of base majority carriers (holes) into emitter. This provides higher current gain for the same base doping level compared to a BJT. This margin is used to increase the doping at the base of an HBT in order to reduce the base resistance and to improve linearity by increasing Early voltage [13]. Additionally, Ge composition is graded across the base in order to create an accelerating electric field for minority carriers (electrons) and thus to achieve higher operating frequency. Carbon (C) is introduced into the highly doped SiGe base to prevent base doping from outdiffusion, contributing to the name SiGe:C of the technology.

3.2.1 HBT Layout and Modeling Considerations

The HBT transistor model implemented in B7HF200 is based on the modified Gummel-Poon model [14] adapted for SPICE simulations [15]. Similarly to the one described in section 3.1.1, this model has to be extended by external layout-dependent parasitics C_{bc}, C_{be}, R_b and R_c, as shown in Fig. 3.5.

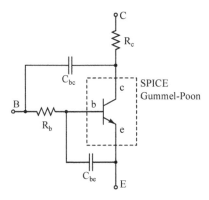

Fig. 3.5 Extended HBT simulation model.

The components C_{bc}, C_{be} describe the parasitic capacitance within the HBT structure, but outside the active area. C_{bc} describes the capacitance between the base contact and the buried layer connection to the collector, whilst C_{be} describes the capacitance due to overlap of base and emitter contacts. In this case the capacitance between the metal interconnects, described in section 3.1.1 is less of concern, since usually only one bottom layer in B7HF200 is sufficient to carry out the required current compared to several thin layers in C11N. Resistances R_b and R_c describe base and collector polysilicon and buried layer interconnect resistances, respectively. Their values depend strongly on the layout configuration.

The Infineon's B7HF200 technology design kit provides a set of standard *npn* layout cells and corresponding extended models for a wide range of emitter mask areas and various contact configurations. Therefore, unlike in C11N, all external parasitics are already defined in this kit, saving time and effort. The choice of a suitable transistor layout geometry is particularly important for high frequency design in order to minimize the parasitics.

According to Fig. 3.4, the standard configuration is base emitter collector (BEC), also depicted on the left in Fig. 3.6. In this configuration, in order to contact both sides of an active region a long bypass of base polysilicon interconnection is re-

3.2 Bipolar Transistors 25

quired, causing higher R_b. In order to reduce R_b the active region can be contacted directly at the second end by an additional base contact. This base emitter base collector (BEBC) configuration is shown on the right in Fig. 3.6.

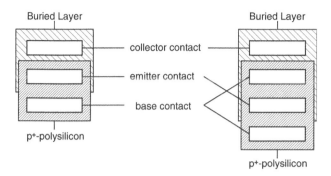

Fig. 3.6 Transistor layout in BEC and BEBC configurations [16].

However, wider overlap between base and collector contacts causes higher base-collector capacitance C_{bc}, thus posing a trade-off between R_b and C_{bc}. The larger horizontal dimension of the transistor results in a wider buried layer and thus higher collector-substrate capacitance C_{cs}. Furthermore, the increased distance between the active transistor area and the collector contact results in an increased R_c. Numerous further transistor configurations, such as e.g. CBEBC, CBEBEBC or BECEB, are available in the design kit. Each configuration offers specific advantages and should be carefully chosen according to the application in order to optimize circuit performance.

3.2.2 Devices Available in B7HF200

Infineon's B7HF200 SiGe:C technology provides vertical *npn* and *pnp* transistors. However, *pnp* devices have a transit frequency f_T of 3.5 GHz and are usually not used for microwave circuit design. The available *npn* transistors are classified with respect to their speed and breakdown voltage into high-voltage (HV), ultra-high-speed (UHS) and high-speed (HS) devices [17]. The latter type is used in this work due to the optimal high-frequency performance.

Several types of resistors are provided. The non-silicided p^+ and p^- polysilicon resistors offer sheet resistance of 150 Ω/□ and 1000 Ω/□, respectively. Furthermore, tantalum-nitride (TaN) high precision metal resistor with sheet resistance of 20 Ω/□ offers a high current carrying capability.

B7HF200 provides a MIM capacitor with a specific capacitance of 1.4 fF/μm². It is integrated in the metallization and uses a 50 nm Al_2O_2 dielectric layer.

3.3 Technology Comparison

In order to evaluate the performance of the available technologies and to verify their suitability for the application, this section summarizes the key parameters.

3.3.1 Transistor Performance

For comparison of high-frequency characteristics offered by Infineon's C11N and B7HF200 transistors two sizes of devices, commonly used throughout this work, have been analyzed. The low threshold voltage NMOS (NLVT) and PMOS (PLVT) transistors are 40 μm wide and have the smallest allowable gate length of 0.13 μm. The SiGe *npn* bipolar transistor has a drawn emitter area of A_E=0.35 × 10 μm^2 and contact configuration CBEBEBC.

One of the main figures of merit used to quantify the transistor high-frequency performance is the maximum oscillation frequency f_{max}, defined under consideration of an optimum matching at input and output ports as the frequency at which the power gain becomes unity. However, for MOS transistors f_{max} strongly depends on gate finger width and contact configuration, whilst it remains almost constant for different HBT sizes [18]. Thus, another common figure of merit, the transit frequency f_T, defined as the frequency at which the current gain becomes unity, is used here for comparison of high-frequency performance of the devices. Simulation of f_T as function of drain-source bias current for MOSFET transistors at $|V_{ds}|$=0.9 V and as function of collector current for HBT at V_{bc}=0 V is presented in Fig. 3.7(a) and Fig. 3.7(b), respectively. Due to SPICE limitations high-current effects are not

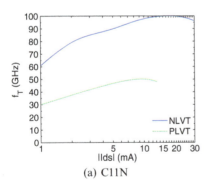
(a) C11N

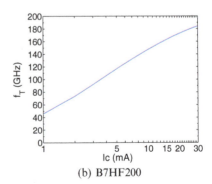
(b) B7HF200

Fig. 3.7 Simulated transit frequency f_T versus bias current.

included in the available HBT model, limiting its validity up to a current of 30 mA or a density of 8 mA/μm^2, which is in the vicinity of the optimum current density for maximum f_T. The maximum transit frequency of 100 GHz, 55 GHz and

3.3 Technology Comparison

180 GHz is obtained at a bias current of 18 mA, 9 mA and 16 mA for NLVT, PLVT and *npn*, respectively. The optimum current densities for peak f_T of 0.45 mA/μm and 0.23 mA/μm for NMOS and PMOS devices, respectively, correspond to the value predicted in the work [19], which suggests invariance of optimal biasing over various technology nodes or foundries.

The PMOS transistor is much slower due to low mobility of holes. The *npn* transistor offers very large margin for realization of 24 GHz applications. This is particularly advantageous for circuit performance under temperature variation, as can be observed in Fig. 3.8 that depicts variation of f_T when transistors are biased for the maximum operating frequency. The transit frequency of CMOS transistors seems

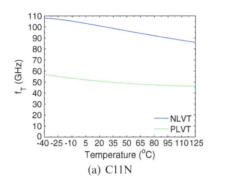
(a) C11N

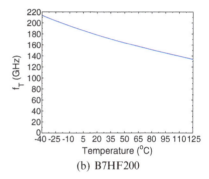
(b) B7HF200

Fig. 3.8 Simulated transit frequency f_T versus temperature.

to exhibit lower temperature variation. Additional degradation of performance parameters other than f_T, occurs in CMOS due to short-channel effects. Furthermore, according to Fig. 3.7 a very high drain current density and a large headroom are required in order to achieve the maximum f_T. Thus, the transistors are usually operated well below the peak operation frequency. Therefore, at high temperatures the bipolar transistor offers higher performance margin.

The optimum bias current for the highest transconductance g_m and thus the highest gain coincides with the optimum current for the highest f_T, as shown for 24 GHz in Fig. 3.9. The bipolar transistor offers a much higher transconductance of 350 mS compared to 28 mS of NMOS transistor for these dimensions at the same bias current of 20 mA. In order to achieve similar peak g_m for NMOSFET, a large device width is required, which results in a higher parasitic capacitance.

For receiver building blocks as LNA and mixer, the transistors may be biased for the lowest minimal achievable noise figure NF_{min}. As can be observed in Fig. 3.10 the optimal current density for CMOS roughly corresponds to the predicted in [19] and is much lower than the current required for the highest g_m. Thus, there is a tradeoff between the transistor highest operation frequency, gain and minimum achievable noise figure. The HBT offers lower minimal noise figure of 1.2 dB compared to 1.6 dB and 2.1dB of N- and PMOSFET, respectively.

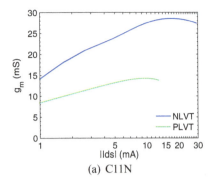
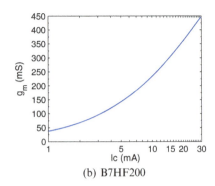

Fig. 3.9 Simulated transconductance g_m at 24 GHz versus bias current.

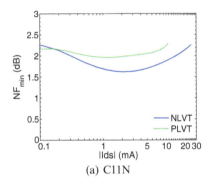
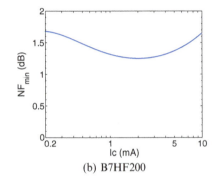

Fig. 3.10 Simulated NF$_{min}$ at 24 GHz versus bias current.

An additional transistor feature, particularly important for direct down-conversion mixers and voltage-controlled oscillators (VCO), is the low frequency noise. It strongly depends on bias current. For typical biasing, a bipolar transistor offers a very low flicker noise corner frequency f_c of some kilohertz [20], whilst CMOS transistors have f_c in the range of several megahertz.

The key parameters of the presented transistors are summarized in Table 3.1.

Table 3.1 Comparison of transistor performance in C11N and B7HF200.

Parameter	NLVT @$|V_{ds}|$=0.9 V	PLVT @$|V_{ds}|$=0.9 V	HS-*npn* @V_{bc}=0 V
Drawn size (μm^2)	0.13 × 40	0.13 × 40	0.35 × 20
Peak f_T (GHz)	100 @$|I_d|$=16 mA	50 @$|I_d|$=9 mA	180 @I_c=28 mA
Peak g_m (mS)	26 @$|I_d|$=16 mA	12 @$|I_d|$=9 mA	440 @I_c=28 mA
Min NF$_{min}$ (dB)	1.6 @$|I_d|$=4 mA	2.1 @$|I_d|$=2 mA	1.2 @I_c=2 mA

3.3.2 Metallization and Passive Components

An additional crucial aspect for comparison of technologies is the available metallization. The cross-sections of the standard back end of line (BEOL), offered by C11N and B7HF200 processes are compared to scale in Fig. 3.11. C11N offers six copper layers M1-M6 and a 1.35 μm thick aluminium top layer, whilst B7HF200 offers four copper layers having a 2.75 μm thick top metal and a pad layer.

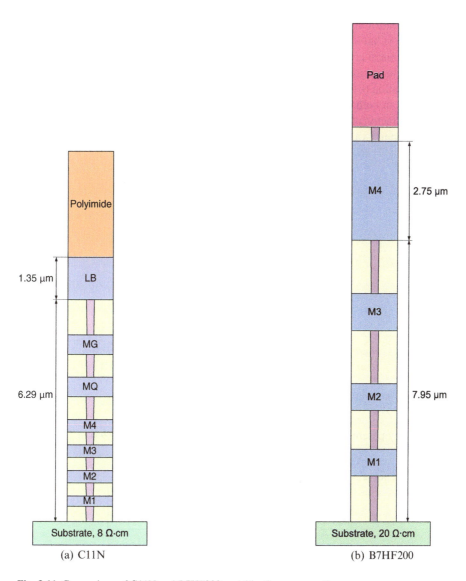

Fig. 3.11 Comparison of C11N and B7HF200 metallization cross-sections.

Typical resistivity of the CMOS and SiGe substrates is 8 Ω·cm and 20 Ω·cm, respectively. Obviously, implementation of a thick top metal results in a higher quality factor of inductors in B7HF200 due to a reduced ohmic series resistance. However, at higher frequencies when the skin effect is dominant, the advantage of a thick top metal is reduced. The current flows in a narrow volume close to the conductor surface defined by the skin depth [21]. Thus, higher sidewalls increase the cross-sectional area to a lesser extent compared to the case when current flows through the whole conductor volume. Whereas, the increased distance to substrate and the higher substrate resistivity in B7HF200 reduce substrate losses and have stronger contribution to the quality factor improvement.

Inductors having 340 pH at 24 GHz realized as octagonal spiral coils with 2 turns in the top metal layer have been simulated in C11N and B7HF200 technologies. The inductors have outer radius of 31.8 μm and 35 μm, inner radius of 18 μm and 21 μm and separation between the traces of 1.8 μm and 2.4 μm, respectively. The width of the traces in both cases is 6 μm. As shown in Fig. 3.12, the inductor in C11N has a quality factor of 12.4 compared to 20.6 of the inductor in B7HF200 at 24 GHz. The maximum value of 12.6 and 21.1 is achieved at 23.5 GHz and 30 GHz for the inductor in C11N and B7HF200, respectively.

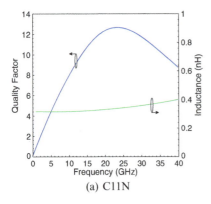
(a) C11N

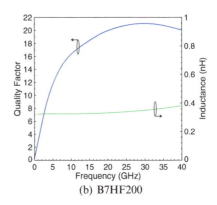
(b) B7HF200

Fig. 3.12 Comparison of inductors.

The key parameters of the available technologies are summarized in Table 3.2.

Table 3.2 Comparison of technology parameters in C11N and B7HF200.

Parameter	C11N	B7HF200
No. routing layers	7	4
Substrate resistivity (Ω·cm)	8	20
Top metal thickness (μm)	1.35	2.75
Top metal distance to substrate (μm)	6.29	7.95

References

1. M. Tiebout, *Low power low phase noise fully integrated VCO design in standard CMOS*, Dissertation, TU Berlin, 2004.
2. T. Schiml and *et al.*, "A 0.13 μm CMOS Platform with Cu/ Low-k Interconnects for System On Chip Applications", *in VLSI Technology Digest of Technical Papers*, pp. 101--102. IEEE, 2001.
3. Infineon, "130 nm CMOS Platform Technology", Technology Brochure, `http://www.infineon.com/cms/en/product/applications/wireless-communication/custom-rf/index.html`, 2001.
4. T. Ytterdal, Y. Cheng, and T. A. Fjeldly, *Device Modeling for Analog and RF CMOS Circuit Design*, Wiley, 2003.
5. H. Shichman and D. A. Hodges, "Modeling and simulation of insulated-gate field-effect transistor switching circuits", *IEEE Journal of Solid-State Circuits*, vol. 3, pp. 285--289, September 1968.
6. B. Razavi, *Design of Analog CMOS Integrated Circuits*, McGraw Hill, 2001.
7. Device Group Berkeley, "Berkeley Short Channel IGFET Model", available at `http://www-device.eecs.berkeley.edu/~bsim3/bsim4.html`, 2006.
8. C. Weyers, D. Kehrer, J. Kunze, P. Mayr, D. Siprak, M. Tiebout, J. Hausner, and U. Langmann, "Improved RF-Performance of Sub-Micron CMOS Transistors by Asymmetrically Fingered Device Layout", *in IEEE Radio Frequency Integrated Circuits (RFIC) Symposium Digest*, pp. 563--566, Atlanta, USA, June 2008.
9. M. J. Deen, C.-H. Chen, S. Asgaran, G. A. Rezvani, J. Tao, and Y. Kiyota, "High-Frequency Noise of Modern MOSFETs: Compact Modeling and Measurement Issues", *IEEE Transactions on Electron Devices*, vol. 53, pp. 2062--2081, September 2006.
10. J. Böck, H. Schäfer, K. Aufinger, R. Stengl, S. Boguth, R. Schreiter, M. Rest, H. Knapp, M. Wurzer, W. Perndl, T. Böttner, and T. F. Meister, "SiGe Bipolar Technology for Automotive Radar Applications", *in Proc. Bipolar/BiCMOS Circuits and Technology (BCTM)*, pp. 84--87, Montreal, Canada, September 2004.
11. Infineon, "B7HF200 200 GHz SiGe", Technology Brochure, `http://www.infineon.com/cms/en/product/applications/wireless-communication/custom-rf/index.html`, 2005.
12. A. Cuthbertson and P. Ashburn, "Self-Aligned Transistors with Polysilicon Emitters for Bipolar VLSI", *IEEE Journal of Solid-State Circuits*, vol. 20, pp. 162--167, February 1985.
13. J.-F. Luy and P. Russer, *Silicon-Based Millimeter-Wave Devices*, Springer Verlag, 1994.
14. H. K. Gummel and H. C. Poon, "An Integral Charge Control Model of Bipolar Transistors", *The Bell System Technical Journal*, vol. 49, pp. 827--852, May 1970.
15. M. Reisch, *High-frequency Bipolar Transistors*, Springer Verlag, 2003.
16. H. Knapp, *Realisierung optimierter monolithisch integrierter Oszillatoren und Frequenzteiler für Mikrowellen in Si- und SiGe-Technologie*, Dissertation, Institut für Nachrichten- und Hochfrequenztechnik der TU Wien, 1999.
17. R. K. Vytla, T. F. Meister, K. Aufinger, D. Lukashevich, S. Boguth, H. Knapp, J. Böck, H. Schäfer, and R. Lachner, "Simultaneous Integration of SiGe High Speed Transistors and High Voltage Transistors", *in Proc. Bipolar / BiCMOS Circuits and Technology Meeting (BCTM)*, pp. 61--64, Maastricht, The Netherlands, September 2006.
18. S. Voinigescu, D. S. McPherson, F. Pera, S. Szilagzyi, M. Tazlauanu, and H. Tran, "Comparison of Silicon and III-V Technology Performance and Building Block Implementations for 10 and 40 Gb/s Optical Networking ICs", *Journal of High Speed Electronics and Systems*, vol. 13, pp. 25--57, March 2003.
19. T. O. Dickson, K. H. K. Yau, T. Chalvatzis, A. M. Mangan, E. Laskin, R. Beerkens, P. Westergaard, M. Tazlauanu, M.-T. Yang, and S. P. Voinigescu, "The Invariance of Characteristic Current Densities in Nanoscale MOSFETs and Its Impact on Algorithmic Design Methodologies and Design Porting of Si(Ge) (Bi)CMOS High-Speed Building Blocks", *IEEE Journal of Solid-State Circuits*, vol. 41, pp. 1830--1845, August 2006.

20. R. Gabl, K. Aufinger, J. Böck, and T. F. Meister, "Low-Frequency Noise Characteristics of Advanced Si and SiGe Bipolar Transistors", *in Proc. of European Solid-State Device Research Conference (ESSDERC)*, pp. 536--539, Stuttgart, Germany, September 1997.
21. D. Pozar, *Microwave Engineering*, Wiley, 2nd edition, 1998.

Chapter 4
Modeling Techniques

Accurate modeling of components is essential for circuit design at microwave frequencies. Particular care is required during the design stage in order to consider the necessary parasitic effects. In the designs throughout this work all the passive components and the on-chip interconnects, including the pads and the transmission lines leading to the active components, are carefully simulated using the 2.5D field solver SonnetEM. The obtained S-parameter models of the metallization are then included in simulations in Agilent's Advanced Design System (ADS). However, using S-parameter models in time-domain simulations often leads to convergence and causality issues. Therefore, in some cases lumped element equivalent circuits are used instead of the frequency based data to model passive on-chip components. This requires techniques to fit accurately frequency dependent S-parameters to an equivalent circuit in a given frequency range.

Additional modeling efforts are required for active devices in CMOS. As mentioned in section 3.1.1, a BSIM model includes only the internal transistor parasitics. Thus, it has to be extended by additional layout-dependent external parasitics describing the capacitance between the metal interconnects leading to the transistor terminals. However, obtaining values of these external capacitances is usually cumbersome and time-consuming.

Therefore, two modeling techniques are proposed in this work. Section 4.1 describes a technique for fast fitting of equivalent circuit values to the given S-parameters data of inductors. Section 4.2 describes a method for fast estimation of external parasitic metallization capacitance of transistor finger layout.

4.1 Analytical Fitting of On-Chip Inductors

A significant amount of research has been focused on modeling of inductors over the past years. The models of inductors are usually obtained in the frequency domain either through measurement or from an electromagnetic field solver simulation in form of an S-parameter file. Numerous works have been published on the develop-

V. Issakov, *Microwave Circuits for 24 GHz Automotive Radar in Silicon-based* 33
Technologies, DOI 10.1007/978-3-642-13598-9_4, © Springer-Verlag Berlin Heidelberg 2010

ment of a physical procedure for broadband equivalent circuit parameters extraction from the given S-parameters frequency data [1], [2]. These approaches provide accurate results, but are usually time-consuming and require additional information about the process and inductor layout. Another option to generate an equivalent lumped element circuit is by data fitting using rational functions. This approach provides the possibility of minimal order circuit synthesis for an arbitrary frequency dependent dataset [3]. However, the resulting circuits are less physically intuitive. An additional approach, commonly used in practice, is to perform numerical optimization by means of a computer aided design (CAD) tool, such as e.g. ADS. This is simple, but requires many iterations that might not converge. The first analytical closed-form expressions for direct non-iterative fitting of measured or simulated S-parameter data of an inductor to a single-π equivalent circuit in the least-square-error sense are reported in [4] and described in detail in this section.

On-chip inductors exhibit physical effects that can be modeled by lumped elements as depicted in Fig. 4.1(a) and combined into an equivalent circuit. The simplest inductor model is the single-π circuit [5] described in Fig. 4.1(b).

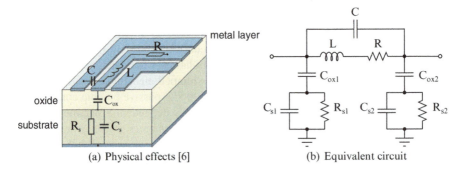

(a) Physical effects [6]   (b) Equivalent circuit

Fig. 4.1 Visualization of lumped element equivalent circuit.

The elements of the equivalent circuit in Fig. 4.1(b) are interpreted as follows:

- L - is the primary inductance of the coil
- R - models the ohmic losses due to the finite conductivity of traces, skin-effect and current crowding
- C - models the capacitive coupling between the windings
- C_{ox1}, C_{ox2} - describe the capacitance between the inductor and the substrate
- C_{s1}, C_{s2} - describe the capacitive effects in the substrate
- R_{s1}, R_{s2} - represent the ohmic losses in the substrate

The lumped element modeling approach is valid since the physical size of the structure is usually much smaller compared to the guided wavelength corresponding to the operating frequency. The simplistic model in Fig. 4.1(b) does not cover all the effects. Some effects as for example eddy currents in conductive substrates, the proximity effect or distributed effects of a large inductor are not accurately described by

4.1 Analytical Fitting of On-Chip Inductors

this model and a more advanced elaborate equivalent circuit representation, such as e.g. in [2], may be required. However, it is impractical to apply analytical fitting to the advanced model. Additionally, for most well-behaving inductors the simple single-π description in Fig. 4.1(b) is usually adequate below the self-resonance frequency (SRF).

Therefore, for simplicity the data have been fitted to the 9-element spiral equivalent circuit in Fig. 4.1. The proposed approach is based on Y-parameters, thus prior to applying the analysis, the S-parameter data have to be converted into Y-parameters. The conversion is given in matrix form for a 50 Ω system using the well-known formula [7]

$$Y = (I - S) \cdot (I + S)^{-1}, \tag{4.1}$$

where I is the unity matrix. The π-shape of the equivalent circuit in Fig. 4.1 is used to simplify the analysis and consider the branches separately. The network can be decomposed into branches using the well-known representation [8]

$$Y = \begin{bmatrix} y_{11} & y_{12} \\ y_{21} & y_{22} \end{bmatrix} = \begin{bmatrix} y_{p1} + y_s & -y_s \\ -y_s & y_s + y_{p2} \end{bmatrix}, \tag{4.2}$$

where y_s and $y_{p1,2}$ describe the admittance of the series branch and the shunt branches at ports 1 and 2, respectively, as shown in Fig. 4.2.

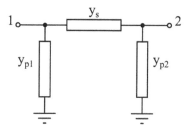

Fig. 4.2 Branch decomposition of a two-port network using Y-parameters.

Eq. (4.2) can be rearranged to obtain the values of the branch admittances

$$y_s(\omega) = -y_{12}(\omega), \tag{4.3}$$
$$y_{p1}(\omega) = y_{11}(\omega) + y_{12}(\omega), \tag{4.4}$$
$$y_{p2}(\omega) = y_{21}(\omega) + y_{22}(\omega). \tag{4.5}$$

This considerably simplifies the problem, since the admittance data of each branch can be used to fit the optimal lumped element values for that branch separately.

4.1.1 Series Branch Parameters Fitting

In the next step, the series branch is considered. The numerical frequency dependent admittance dataset is calculated by (4.3). The analytical expression of the series branch admittance as a function of angular frequency ω is given by

$$y_s(\omega) = \frac{R}{R^2 + \omega^2 L^2} + j\omega \left(C - \frac{L}{R^2 + \omega^2 L^2} \right). \tag{4.6}$$

This expression is used for fitting of the optimal values for R, L and C in Fig. 4.1(b) in the least-square error (LSE) sense [9]. Depending on the required frequency range, in which the data have to be fitted, a sub-dataset, consisting of N frequency-admittance points, shall be extracted from y_s. The ith admittance point in the given dataset is denoted as $y_i = -y_{21}(\omega_i) = y_s(\omega_i)$.

The reciprocal of the real part of admittance in (4.6) is used to fit the resistance and inductance of the series branch in the least square error sense. The squared error is thus given by

$$S(R,L) = \sum_{i=1}^{N} \left(\frac{R^2 + \omega_i^2 L^2}{R} - \frac{1}{\text{Re}\{y_i\}} \right)^2. \tag{4.7}$$

The error should be minimized, thus the extremum of the function is determined by applying derivatives over R and L and equating both to zero

$$\frac{\partial S}{\partial R} = 2 \sum_{i=1}^{N} \left(R + \frac{L^2}{R}\omega_i^2 - \frac{1}{\text{Re}\{y_i\}} \right) \left(1 - \frac{L^2}{R^2}\omega_i^2 \right) = 0$$

$$\frac{\partial S}{\partial L} = 2 \sum_{i=1}^{N} \left(R + \frac{L^2}{R}\omega_i^2 - \frac{1}{\text{Re}\{y_i\}} \right) \left(\frac{2L}{R}\omega_i^2 \right) = 0. \tag{4.8}$$

Further simplification of the above system of equations leads to

$$N \cdot R - \left(\sum_{i=1}^{N} \omega_i^4 \right) \cdot \frac{L^4}{R^3} + \left(\sum_{i=1}^{N} \frac{\omega_i^2}{\text{Re}\{y_i\}} \right) \cdot \frac{L^2}{R^2} - \left(\sum_{i=1}^{N} \frac{1}{\text{Re}\{y_i\}} \right) = 0$$

$$\left(\sum_{i=1}^{N} \omega_i^2 \right) \cdot R + \left(\sum_{i=1}^{N} \omega_i^4 \right) \cdot \frac{L^2}{R} - \left(\sum_{i=1}^{N} \frac{\omega_i^2}{\text{Re}\{y_i\}} \right) = 0. \tag{4.9}$$

Second equation in (4.9) is multiplied by $(L/R)^2$ and added to the first, giving

$$\left(\sum_{i=1}^{N} \omega_i^2 \right) \cdot \frac{L^2}{R} + N \cdot R = \sum_{i=1}^{N} \frac{1}{\text{Re}\{y_i\}} \Rightarrow \frac{L^2}{R} = \frac{\left(\sum_{i=1}^{N} \frac{1}{\text{Re}\{y_i\}} \right)}{\left(\sum_{i=1}^{N} \omega_i^2 \right)} - \frac{N}{\left(\sum_{i=1}^{N} \omega_i^2 \right)} \cdot R. \tag{4.10}$$

This is then substituted into the second equation of (4.9) and the optimal series resistance R that fits the dataset y_s in the frequency range ω_1 to ω_N is given by

4.1 Analytical Fitting of On-Chip Inductors

$$R = \frac{\left(\sum\limits_{i=1}^{N} \frac{1}{\mathrm{Re}\{y_i\}}\right) \cdot \left(\sum\limits_{i=1}^{N} \omega_i^4\right) - \left(\sum\limits_{i=1}^{N} \frac{\omega_i^2}{\mathrm{Re}\{y_i\}}\right) \cdot \left(\sum\limits_{i=1}^{N} \omega_i^2\right)}{N \cdot \left(\sum\limits_{i=1}^{N} \omega_i^4\right) - \left(\sum\limits_{i=1}^{N} \omega_i^2\right)^2}. \tag{4.11}$$

The optimal inductance is easily obtained by plugging (4.11) into (4.10)

$$L = \left(\frac{\left(\sum\limits_{i=1}^{N} \frac{1}{\mathrm{Re}\{y_i\}}\right) - R \cdot N}{\left(\sum\limits_{i=1}^{N} \omega_i^2\right)} \cdot R\right)^{1/2}. \tag{4.12}$$

Now in the similar manner the imaginary part of the series branch admittance in (4.6) shall be used to fit the capacitance between the windings

$$S(C) = \sum\limits_{i=1}^{N} \left(\omega_i C - \frac{\omega_i L}{R^2 + \omega_i^2 L^2} - \mathrm{Im}\{y_i\}\right)^2. \tag{4.13}$$

In order to find the minimum error, a derivative is applied

$$\frac{\partial S}{\partial C} = 2 \sum\limits_{i=1}^{N} \left(\omega_i^2 C - \frac{\omega_i^2 L}{R^2 + \omega_i^2 L^2} - \omega_i \mathrm{Im}\{y_i\}\right) = 0. \tag{4.14}$$

Thus, the optimal capacitance is given by

$$C = \frac{\sum\limits_{i=1}^{N} \frac{\omega_i^2 L}{R^2 + \omega_i^2 L^2} + \sum\limits_{i=1}^{N} \omega_i \mathrm{Im}\{y_i\}}{\sum\limits_{i=1}^{N} \omega_i^2}, \tag{4.15}$$

where ω_i is i^{th} frequency point and N is the number of points in the dataset.

However, an accurate observation of the presented analytical derivation reveals that the dataset has to fulfill several conditions for this approach to be applicable. From (4.12) for the inductance value L to be real and non-zero must exist

$$\frac{\sum\limits_{i=1}^{N} \frac{1}{\mathrm{Re}\{y_i\}}}{N} > R. \tag{4.16}$$

This implies that the average of the real part values of series branch admittance over the whole dataset must be greater than the calculated optimal value of R. In case that the data in the dataset do not fulfill this condition, this means that the estimation of R in the least square error sense is higher than reflected by the simple equivalent circuit model having constant component values, due to additional frequency-dependent loss mechanisms.

38 4 Modeling Techniques

Additionally, it has to be emphasized that in order to be able to resolve this analytically, the real part of the admittance has been treated separately and first of all R and L have been fitted in the optimal way. But if there are additional losses that are not covered, this might also affect the fitting of C. Therefore, this approach is more suitable for "well-behaving" spiral inductors that can be accurately modeled by a single-π equivalent circuit.

4.1.2 Shunt Branches Parameters Fitting

Next, the admittance of the shunt branches is considered. The following analysis is applicable to both branches, due to their identical structure. Thus, the branches are not distinguished in this analysis for generality. In order to calculate the component values for the left branch C_{ox1}, R_{s1} and C_{s1}, the numerical dataset y_{p1} obtained from (4.4) is used, whilst for calculation of C_{ox2}, R_{s2} and C_{s2} the dataset y_{p2} obtained from (4.5) is used. The analytical expression of a shunt branch impedance as a function of angular frequency ω is given by

$$z_p(\omega) = \frac{R_s}{1 + \omega^2 C_s^2 R_s^2} - j\left(\frac{1}{\omega C_{ox}} + \frac{\omega C_s R_s^2}{1 + \omega^2 C_s^2 R_s^2}\right). \tag{4.17}$$

As previously, a dataset consisting of N frequency-admittance points, corresponding to the frequency range of interest, is extracted from y_{p1} and y_{p2} and converted to impedance data. The ith impedance point is denoted as z_i. It is defined for the left hand branch as $z_i = z_{p1}(\omega_i) = 1/y_{p1}(\omega_i)$, whilst for the right hand branch as $z_i = z_{p2}(\omega_i) = 1/y_{p2}(\omega_i)$.

The reciprocal of the real part of impedance in (4.17) is used to fit the capacitance and resistance of the shunt branch in the least square error sense

$$S(R_s, C_s) = \sum_{i=1}^{N} \left(\frac{1 + \omega_i^2 R_s^2 C_s^2}{R_s} - \frac{1}{\operatorname{Re}\{z_i\}}\right)^2. \tag{4.18}$$

The error shall be minimized, thus the extremum of the function in (4.18) is determined by applying derivatives over R_s and C_s and equating both to zero

$$\begin{aligned}
\frac{\partial S}{\partial R_s} &= 2\sum_{i=1}^{N} \left(\frac{1}{R} + \omega_i^2 RC_s^2 - \frac{1}{\operatorname{Re}\{z_i\}}\right)\left(-\frac{1}{R^2} + \omega_i^2 C_s^2\right) = 0 \\
\frac{\partial S}{\partial C_s} &= 2\sum_{i=1}^{N} \left(\frac{1}{R} + \omega_i^2 RC_s^2 - \frac{1}{\operatorname{Re}\{z_i\}}\right) 2\left(\omega_i^2 RC_s\right) = 0.
\end{aligned} \tag{4.19}$$

Further simplification of the above system of equations leads to

4.1 Analytical Fitting of On-Chip Inductors

$$
-\frac{N}{R_s^3} + \left(\sum_{i=1}^{N} \frac{1}{\text{Re}\{z_i\}}\right)\frac{1}{R_s^2} + \left(\sum_{i=1}^{N} \omega_i^4\right)R_s C_s^4 - \left(\sum_{i=1}^{N} \frac{\omega_i^2}{\text{Re}\{z_i\}}\right)C_s^2 = 0
$$

$$
\left(\sum_{i=1}^{N} \omega_i^2\right)\frac{1}{R_s} + \left(\sum_{i=1}^{N} \omega_i^4\right)R_s C_s^2 - \left(\sum_{i=1}^{N} \frac{\omega_i^2}{\text{Re}\{z_i\}}\right) = 0. \tag{4.20}
$$

The above system of equations (4.20) can be easily solved by rearranging the second equation for C_s^2. Then an additional expression for C_s^2 is obtained by multiplying the second equation in (4.20) by C_s^2 and subtracting from the first one. Comparing both expressions results in the following equation

$$
C_s^2 = \frac{\left(\sum_{i=1}^{N} \frac{1}{\text{Re}\{z_i\}}\right)R_s - N}{\left(\sum_{i=1}^{N} \omega_i^2\right)R_s^2} = \frac{\left(\sum_{i=1}^{N} \frac{\omega_i^2}{\text{Re}\{z_i\}}\right)R_s - \left(\sum_{i=1}^{N} \omega_i^2\right)}{\left(\sum_{i=1}^{N} \omega_i^4\right)R_s^2}. \tag{4.21}
$$

Solving the equation for R_s provides the optimal substrate resistance of a shunt branch

$$
R_s = \frac{N\left(\sum_{i=1}^{N} \omega_i^4\right) - \left(\sum_{i=1}^{N} \omega_i^2\right)^2}{\left(\sum_{i=1}^{N} \frac{1}{\text{Re}\{z_i\}}\right)\left(\sum_{i=1}^{N} \omega_i^4\right) - \left(\sum_{i=1}^{N} \frac{\omega_i^2}{\text{Re}\{z_i\}}\right)\left(\sum_{i=1}^{N} \omega_i^2\right)}. \tag{4.22}
$$

Once the value of R_s is obtained, it can be plugged into (4.21) to obtain C_s

$$
C_s = \left(\frac{\left(\sum_{i=1}^{N} \frac{1}{\text{Re}\{z_i\}}\right)R_s - N}{\left(\sum_{i=1}^{N} \omega_i^2\right)R_s^2}\right)^{1/2}. \tag{4.23}
$$

The imaginary part of the shunt branch impedance is used to fit the oxide capacitance

$$
S(C_{\text{ox}}) = \sum_{i=1}^{N} \left(\frac{1}{\omega_i C_{\text{ox}}} + \frac{\omega_i C_s R_s^2}{1 + \omega_i^2 C_s^2 R_s^2} + \text{Im}\{z_i\}\right)^2. \tag{4.24}
$$

Again, derivative is applied to the error function to find an extremum

$$
\frac{\partial S}{\partial C_{\text{ox}}} = \frac{-2}{C_{\text{ox}}^2}\sum_{i=1}^{N} \left(\frac{1}{\omega_i C_{\text{ox}}} + \frac{\omega_i C_s R_s^2}{1 + \omega_i^2 C_s^2 R_s^2} + \text{Im}\{z_i\}\right)\frac{1}{\omega_i} = 0. \tag{4.25}
$$

Thus, the optimal oxide capacitance is given by

$$
C_{\text{ox}} = -\frac{\sum_{i=1}^{N} \frac{1}{\omega_i^2}}{\sum_{i=1}^{N} \frac{R_s^2 C_s}{1 + \omega_i^2 R_s^2 C_s^2} + \sum_{i=1}^{N} \frac{\text{Im}\{z_i\}}{\omega_i^2}}. \tag{4.26}
$$

By observing the above expressions, one can see that in order to have physical element values again several conditions on the dataset have to be fulfilled. From (4.21), in order for the capacitance C_s to be real and non-zero must exist

$$R_s > \frac{N}{\sum_{i=1}^{N} \frac{1}{\text{Re}\{z_i\}}}. \qquad (4.27)$$

This is very similar to (4.16) and implies that in case this condition is not fulfilled, the estimation of shunt losses is higher than described by the simple equivalent circuit. Additional substrate loss effects, such as e.g. eddy-currents are not covered by the simplistic model.

4.1.3 Results Verification

The above analytical results have been verified on measured S-parameters of an on-chip inductor realized in Infineon's C11N technology. The inductor is realized as an octagonal symmetrical spiral coil with six turns on the four lowest 0.35 µm thick copper layers. The outer diameter of the inductor is 143 µm, the inner diameter 65 µm, the width of the outer trace is 7.5 µm and the separation between the traces is 1.9 µm. A micrograph of the inductor is presented in Fig. 4.3. The test structure is designed for a low-frequency inductance of 4.2 nH.

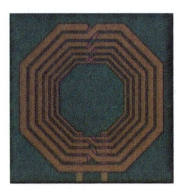

Fig. 4.3 Micrograph of a 4.2 nH inductor in CMOS.

It has been measured on-wafer using Cascade Microtech's Infinity GSG probes and Agilent 8510C VNA. The inductor S-parameters have been de-embedded using the Open-Thru approach [10]. The S-parameters have been converted to Y-parameters and separated into datasets corresponding to the series and shunt branches using (4.3)-(4.5). The optimal values of the lumped element equivalent

4.1 Analytical Fitting of On-Chip Inductors   41

circuit are obtained in the range of 0.1 – 1.4 GHz by applying (4.11), (4.12), (4.15) for the series branch and (4.22), (4.23), (4.26) for the shunt branch, respectively.

The component values are summarized in Table 4.1. These data fulfill the requirements (4.16) and (4.27), since in this frequency range the inductor can be well described using the single-π equivalent circuit.

Table 4.1 Calculated optimal values for a 4.2 nH inductor in CMOS.

Parameter	R	L	C	R_{s1}	C_{s1}	C_{ox1}	R_{s2}	C_{s2}	C_{ox2}
Units	Ω	nH	fF	Ω	pF	fF	Ω	pF	fF
Value	8.43	4.1	108	492	0.44	221	2860	0.35	784

A comparison of the measured S-parameter data with the frequency characteristics of the fitted equivalent circuit is presented in Fig. 4.4.

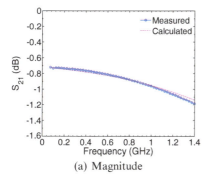
(a) Magnitude

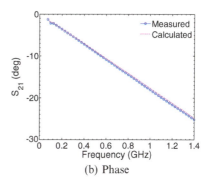
(b) Phase

Fig. 4.4 Comparison of S-parameters for a 4.2 nH inductor.

Additionally, the main inductor parameters, evaluated from the measured and fitted inductor data, are compared in Fig. 4.5. As can be observed, the equivalent circuit achieves an accurate fit to the inductor behavior.

The presented analysis provides simple analytical expressions for fitting frequency dependent S-parameter data of on-chip inductors in the LSE sense. The optimal values of an equivalent single-π circuit are obtained in a direct non-iterative manner. This can be useful for a quick insight into the main equivalent parameters of a spiral inductor in a certain frequency range. However, due to limitations of the 9-element inductor model, the fitting procedure might provide non-physical values. Using an advanced model, would provide robust fitting, but expressions for fitting the data in the LSE sense cannot be derived analytically. In that case a physical approach [2] or numerical optimization procedure can be used.

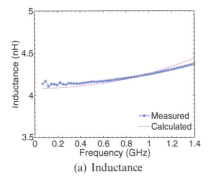
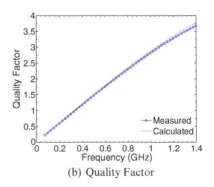

(a) Inductance (b) Quality Factor

Fig. 4.5 Comparison of quality factor and inductance for a 4.2 nH inductor.

4.2 Transistor Finger Capacitance Estimation

Accurate estimation of the external parasitic capacitances due to transistor finger metallization, necessary for extending a BSIM model for CMOS transistors as explained in detail in section 3.1.1, is a challenge. The capacitance values are usually evaluated either by means of 3D electromagnetic field-solver simulations or through layout parasitic extraction in the design environment. The first option is very accurate, but time-consuming. The second option is fast, but less accurate. An additional option, analyzed in this section, provides a fast and accurate capacitance estimation of a 3D structure without the use of a field solver. It is proposed to use the surface charge method (SCM) to estimate the transistor finger metallization capacitance numerically. Due to simplicity of the layout geometry, simple parametric meshing and fast numerical solution is possible.

The surface charge method is widely used in different applications in electrostatics, such as e.g. high voltage insulation systems [11], electron optics [12] or biomedicine [13]. However, application of the SCM in microwave and millimeter-wave engineering is uncommon, since most problems require quasi-static or full-wave numerical approaches. The latter are implemented in numerous commercial electromagnetic field solvers. As explained further in this section, an electrostatic approximation offered by the surface charge method is sufficiently accurate in this case, due to very small physical dimensions of the analyzed transistor finger structures compared to the wavelength.

The SCM method [14] is based on the fact that in a system of conductors, when potentials are applied, the surface charges arrange themselves to make the interiors of conducting bodies field-free. By means of equivalent charge interaction, the electric potential and field distribution, induced by contributions from virtual surface point charges are derived numerically. The values of these point charges represent the surface charge distribution, which is assumed to be uniform over the rectangular or triangular unit elements obtained by subdividing the surface of an object. For sufficiently dense discretization this procedure allows to calculate the electric

4.2 Transistor Finger Capacitance Estimation

field, potential and capacitances accurately. Several works consider the accuracy of the surface charge method [15], [16]. The capacitance value converges fast with increasing mesh density [17]. Thus, a moderately dense meshing may provide sufficiently accurate results for circuit simulations. The theoretical background of the surface charge method is described in detail in Appendix C. The applicability of the approach has been verified by comparison of the calculated results with field solver simulations.

The analyzed test structure, presented in Fig. 4.6, describes a typical multifinger layout metallization of a CMOS transistor. The transistor is 40 μm wide and is divided into 15 fingers. The physical parameters of the metal and dielectric layers used in this example correspond to the C11N stack-up described in section 3.3.2. Source and drain fingers are realized in the three lower layers M1-M3. The source and drain are connected on layers M2 and M3 on the opposite sides. The layers are connected using via arrays. The polysilicon transistor gate fingers (red) are connected through via arrays to metal rails in layer M1 (green).

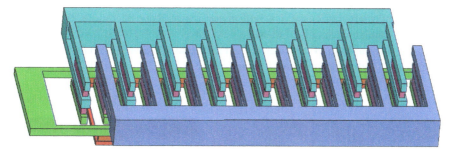

Fig. 4.6 Finger layout of a 40 μm wide transistor in CMOS.

For faster simulation or numerical calculation, the structure has been simplified, i.e. via arrays have been joined into via bars. This structure simplification approach is also common in simulations using commercial 3D field solvers in order to reduce numerical efforts. Additionally, as shown further in Fig. 4.8, the electric field is mainly concentrated between the adjacent faces of the structures. Thus, the main capacitance contribution is due to interaction of the faces of a single cell. Therefore, for a quick approximation only a single cell is considered and the capacitance is multiplied by the amount of cells. Rectangular meshing can be applied in this case for simplicity, since only rectangular faces are present in the structure. Positions of equivalent point charges at the centers of the discretized cells are defined easily by triples of coordinates given in Appendix C.2. The discretized cell with equivalent charges replaced by dots is shown in Fig. 4.7.

Obviously, the equivalent capacitance of the structure varies with frequency. However, the physical size of the transistor finger metallization is considerably smaller than the wavelength of signals in the typical IC applications. For instance, at 24 GHz the on-chip wavelength in dielectric is about 6.3 mm, whilst a typical

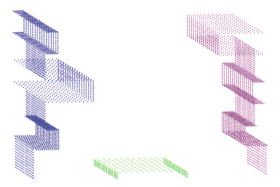

Fig. 4.7 Simplification and discretization of the problem.

finger width is about 3 μm. The dimensions of the analyzed structure are thus much smaller than of any on-chip waveguide. Therefore, distributed and dispersive effects may be neglected in this case. To confirm the presented considerations, the structure has been simulated using the full-wave field solver Ansoft HFSS. As can be seen in Fig. 4.8, the variation of the electric field distribution between 0.1 GHz to 24 GHz is minor. Therefore, a static value obtained using the SCM is a valid approximation.

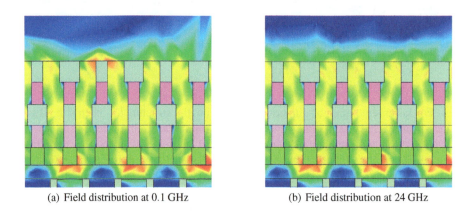

(a) Field distribution at 0.1 GHz   (b) Field distribution at 24 GHz

Fig. 4.8 Complex electric field distribution comparison.

The gate-source C_{gsm}, gate-drain C_{gdm} and drain-source C_{dsm} parasitic capacitances required for the extended transistor model described in Fig. 3.2, are estimated for the given example numerically using the surface charge method and compared with simulated results obtained using the quasistatic field solver Ansoft Q3D in Table 4.2.

Table 4.2 Numerically calculated and simulated capacitance for a 40 μm layout.

Capacitance	C_{gdm} (fF)	C_{gsm} (fF)	C_{dsm} (fF)
Calculated	9.8	2.93	3.04
Simulated	9.72	2.87	2.95

The SCM approach is applicable for quick estimation of transistor finger metallization capacitances. The simplicity of the rectangular problem allows straightforward meshing and simple analytical definition of equivalent charge coordinates. This can be implemented in any programming environment, as e.g. Excel or Matlab, allowing a quick capacitance approximation of any similar finger metallization structure with different physical dimensions. However, if the actual layout deviates from the classical multifinger layout and additional 3D conducting objects have to be taken into account, the code has to be modified or rewritten. Furthermore, the complexity of the manual surface mesh generation and numerical efforts increase significantly with inclusion of any further bodies or problem details into the model.

References

1. M. Kang, J. Gil, and H. Shin, "A Simple Parameter Extraction Method of Spiral On-Chip Inductors", *IEEE Transactions on Electron Devices*, vol. 52, pp. 1976--1981, September 2005.
2. Y. Cao, R. A. Groves, X. Huang, N. D. Zamdmer, J.-O. Plouchart, R. A. Wachnik, T.-J. King, and C. Hu, "Frequency-Independent Equivalent-Circuit Model for On-Chip Spiral Inductors", *IEEE Journal of Solid-State Circuits*, vol. 38, pp. 419--426, March 2003.
3. R. Neumayer, A. Stelzer, F. Haslinger, and R. Weigel, "On the Synthesis of Equivalent-Circuit Models for Multiports Characterized by Frequency-Dependent Parameters", *IEEE Transactions on Microwave Theory and Techniques*, vol. 50, pp. 2789--2796, December 2002.
4. V. Issakov, A. Thiede, M. Wojnowski, K. Büyüktas, and W. Simbürger, "Fast Analytical Parameters Fitting of Planar Spiral Inductors", *in IEEE Conference on Microwaves, Communications, Antennas and Electronic Systems (COMCAS)*, pp. 1--10, Tel Aviv, Israel, May 2008.
5. C. P. Yue and S. S. Wong, "Physical modeling of spiral inductors on silicon", *IEEE Transactions on Electron Devices*, vol. 47, pp. 560--568, March 2000.
6. R. Thüringer, *An Integrated 17 GHz Transmitter in 0.13 μm CMOS for Wireless Applications*, Dissertation, Institut für Nachrichten- und Hochfrequenztechnik der TU Wien, 2005.
7. O. Zinke and H. Brunswig, *Hochfrequenztechnik 1*, chapter 4.11, Springer Verlag, 6th edition, 2000.
8. D. Pozar, *Microwave Engineering*, Wiley, 2nd edition, 1998.
9. J. F. Epperson, *An introduction to numerical methods and analysis*, chapter 4.10, Wiley, 2007.
10. J. Tao, P. Findley, and G.A. Rezvani, "Novel realistic short structure construction for parasitic resistance de-embedding and on-wafer inductor characterization", *in Proc. of the IEEE Conf. on Microelectronic Test Structures*, pp. 187--190, San Jose, USA, April 2005.
11. T. Misaki, H. Tsuboi, K. Itaka, and T. Hara, "Computation of Three-Dimensional Electric Field Problems by a Surface Charge Method and its Application to Optimum Insulator De-

sign", *IEEE Transactions on Power Apparatus and Systems*, vol. PAS-101, pp. 627--634, March 1982.

12. F. H. Read, A. Adams, and J. R. Soto-Montiel, "Electrostatic Cylinder Lenses I: Two-element Lenses", *Journal of Physics E: Scientific Instruments*, vol. 4, pp. 625--632, September 1971.

13. O. Fujiwara and T. Ikawa, "Numerical Calculation of Human-Body Capacitance by Surface Charge Method", *Electronics and Communications in Japan, Part 1*, vol. 85, pp. 38--44, 2002.

14. D. K. Reitan and T. J. Higgins, "Electrical capacitance of the unit cube", *Journal of Applied Physics*, vol. 22, pp. 223--226, February 1951.

15. F. H. Read and N. J. Bowring, "Ultimate numerical accuracy of the surface charge method for electrostatics", *in International Conference on Computation in Electromagnetics (CEM)*, pp. 57--61, Bath, UK, April 1996. IEE.

16. A. Tatematsu, S. Hamada, T. Takuma, and H. Morii, "A study on the accuracy of surface charge measurement", *IEEE Transactions on Dielectrics and Electrical Insulation*, vol. 9, pp. 406--415, June 2002.

17. E. Goto, Y. Shi, and N. Yoshida, "Extrapolated surface charge method for capacity calculation of polygons and polyhedra", *Journal of computational physics*, vol. 100, pp. 105--115, 1992.

Chapter 5
Measurement Techniques

Accurate measurements are essential in high-frequency engineering. Depending on the functionality of the device under test (DUT), different types of measurements are required. Linear characteristics of microwave circuits and devices are usually measured using a vector network analyzer (VNA). The obtained S-parameters are used to describe the main performance characteristics of the networks operating in their linear range. For example, for an LNA the small-signal S-parameters describe its gain, reverse isolation and port matching. Also for instance, the main properties of a directional coupler such as e.g. amplitude and phase imbalance, insertion loss, isolation and port matching can be derived from its S-parameters.

For non-linear characterization of microwave circuits diverse measurement instruments are required. Many non-linear as well as linear properties can be measured using a spectrum analyzer (SPA) in combination with other instruments as e.g. signal generator or noise source. For an LNA these properties are e.g. noise figure and linearity. For a power amplifier (PA) these properties are e.g. the maximum output power and linearity. For frequency-converting circuits, such e.g. down-conversion mixers or receivers, the main properties include conversion gain, noise figure and linearity. For frequency generation circuits, such as e.g. VCO, these are output power, tuning range, output spectrum and phase noise. Obviously, several measurements can be performed using different instruments. For example, noise figure of an LNA or a mixer can be also measured using a noise figure meter (NFM). A brief summary of measurement methods is given in Appendix D.

In all the microwave measurements there are physical components between the ports of the measurement equipment and the ports of the DUT. These components are considered to be the error networks and their impact has to be removed from measured values in order to obtain the actual DUT characteristics. In S-parameter measurements this is done by classical calibration techniques, which are typically implemented in the software package of most VNAs. However, it is often not possible to set the measurement reference planes directly at the ports of the DUT. Thus, the impact of an error network between the calibration reference plane and the DUT has to be removed using any de-embedding technique.

V. Issakov, *Microwave Circuits for 24 GHz Automotive Radar in Silicon-based Technologies*, DOI 10.1007/978-3-642-13598-9_5, © Springer-Verlag Berlin Heidelberg 2010

For characterization of active DUTs using an SPA or NFM, the insertion loss of the error networks in the measurement setup must be determined. The magnitude of the insertion loss is then used in the equations for the cascaded connection of the DUT and the error networks in order to remove the impact of the error networks from the measured parameters and obtain the actual DUT properties.

Differential signaling becomes increasingly popular in active circuits due to its superior noise immunity and ground bounce insensitivity. However, characterization of differential circuits at microwave frequencies is a challenge. Due to cost considerations a commercial four-port or a true-differential VNA might not be easily available. Therefore, very often baluns are introduced and a differential DUT is characterized using single-ended equipment. This approach is valid only if the "balun" exhibits nearly ideal characteristics. However, there is little information in the literature on the accuracy of de-embedding using non-ideal baluns.

This chapter describes several measurement considerations, developed during characterization of passive and active circuits in this work. Section 5.1 presents the de-embedding techniques proposed in this work for characterization of asymmetrical and differential devices. Section 5.2 analyzes thoroughly the measurement accuracy of differential devices using a two-port VNA and baluns.

5.1 S-parameter De-embedding Techniques

In S-parameter measurements, the reference planes are defined by means of a standard calibration technique such as Short-Open-Load-Thru (SOLT), Line-Reflect-Reflect-Match (LRRM) or Thru-Reflect-Line (TRL). A detailed overview of the classical calibration methods is given in [1].

In the case that it is not possible to set the reference planes directly at the measured device, the classical de-embedding techniques such as e.g. Open-Short [2], Pad-Open-Short [3] or Thru [4] have to be applied to remove the impact of any error network between the reference plane and the measured device. The standard on-wafer de-embedding techniques, compared in detail in [5], can be divided into two categories.

The first category consists of techniques based on equivalent lumped-element circuit models such as Short-Open, the three step or the four step methods. These approaches assume a specific lumped-element model of interconnects. This reduces the de-embedding accuracy at higher frequencies.

The second category consists of cascade-based two-port techniques, such as e.g. Thru [6], Thru-Line (TL) [7] or TRL [8]. These techniques allow de-embedding to be performed without modeling of the internal structure of the error network. Thus, they are applicable up to higher frequencies and offer better accuracy than the equivalent circuit based techniques. This section focuses on the cascade-based de-embedding techniques and proposes several extensions.

5.1.1 Extension of Thru Technique for De-embedding of Asymmetrical Error Networks

The cascade-based Thru approach offers the advantage of simplicity, since it uses only one test structure: a symmetrical thru. This also saves the chip or board space and therefore reduces measurement costs. However, this method requires certain restricting assumptions, since the error network has several unknown terms that cannot be resolved in a single measurement. The extension introduced in [9] resolves this disadvantage at the expense of one more measurement and allows accurate characterization of asymmetrical error networks.

5.1.1.1 Theory

Measurement performed with a calibrated vector network analyzer (VNA) provides the S-parameters of a chain connection of networks as shown in Fig. 5.1. The networks A and B are the error networks to be de-embedded.

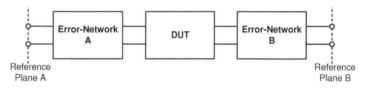

Fig. 5.1 Measured chain connection of DUT and error networks.

It is assumed that the error network B represents the mirrored version of the network A, and that both networks are uncoupled. These conditions are often fulfilled in integrated circuits and precise laminate-based technologies. Using this assumption the expected S-parameters of the thru de-embedding test structure, referred to as α and illustrated in Fig. 5.2, are derived analytically. The S-parameters of the

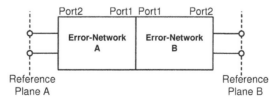

Fig. 5.2 Thru de-embedding test structure α.

error network A can be converted into T-parameters using the well-known relation

$$T_A = \frac{1}{s_{21}} \begin{bmatrix} s_{12}s_{21} - s_{11}s_{22} & s_{11} \\ -s_{22} & 1 \end{bmatrix}. \tag{5.1}$$

When the network is mirrored, ports 2 and 1 are swapped. Thus, the T-parameters of the mirrored error network can be described by

$$T_B = \frac{1}{s_{12}} \begin{bmatrix} s_{12}s_{21} - s_{11}s_{22} & s_{22} \\ -s_{11} & 1 \end{bmatrix}. \tag{5.2}$$

The expected total T-matrix of the thru test structure α is just the product of the T-matrices corresponding to the error boxes A and B, given by (5.1) and (5.2)

$$T_\alpha = \begin{bmatrix} \frac{s_{22}^2-1}{s_{12}s_{21}}s_{11}^2 - 2s_{11}s_{22} + s_{12}s_{21} & -\frac{s_{22}^2-1}{s_{12}s_{21}}s_{11} + s_{22} \\ \frac{s_{22}^2-1}{s_{12}s_{21}}s_{11} - s_{22} & -\frac{s_{22}^2-1}{s_{12}s_{21}} \end{bmatrix}. \tag{5.3}$$

The S-parameter matrix of the thru connection that is expected to be measured is obtained by converting (5.3) back into S-parameters

$$S_\alpha = \begin{bmatrix} s_{11,\alpha} & s_{12,\alpha} \\ s_{21,\alpha} & s_{22,\alpha} \end{bmatrix} = \begin{bmatrix} s_{11} - \frac{s_{22}s_{12}s_{21}}{s_{22}^2-1} & -\frac{s_{12}s_{21}}{s_{22}^2-1} \\ -\frac{s_{12}s_{21}}{s_{22}^2-1} & s_{11} - \frac{s_{22}s_{12}s_{21}}{s_{22}^2-1} \end{bmatrix}. \tag{5.4}$$

As can be observed, the parameters $s_{11,\alpha}$ and $s_{22,\alpha}$ of the S_α matrix are expected to be equal. Similarly, the $s_{12,\alpha}$ and $s_{21,\alpha}$ entries are equal. This means that from a single measurement of a thru structure one gets a system of two independent complex equations and four unknown complex variables

$$s_{11} - \frac{s_{22}s_{12}s_{21}}{s_{22}^2 - 1} = s_{11,\alpha}$$
$$-\frac{s_{12}s_{21}}{s_{22}^2 - 1} = s_{12,\alpha}. \tag{5.5}$$

In order to solve (5.5), additional assumptions are required to reduce the number of unknowns to two. Since the error networks normally do not contain any active devices or ferrites, the error box reciprocity can be assumed

$$s_{12} = s_{21}. \tag{5.6}$$

A further requirement is more restrictive. The Thru method requires that both ports have identical impedance conditions

$$s_{11} = s_{22}. \tag{5.7}$$

Under these assumptions the system of equations (5.5) is simplified to the following form

5.1 S-parameter De-embedding Techniques

$$s_{11}\left(1 - \frac{s_{12}^2}{s_{11}^2 - 1}\right) = s_{11,\alpha}$$
$$-\frac{s_{12}^2}{s_{11}^2 - 1} = s_{12,\alpha}. \tag{5.8}$$

Solution of (5.8) is straightforward

$$s_{11} = s_{22} = \frac{s_{11,\alpha}}{1 + s_{12,\alpha}},$$
$$s_{12} = s_{21} = \pm\sqrt{s_{12,\alpha}\left(1 - s_{11}^2\right)}. \tag{5.9}$$

The condition (5.7) is usually hard to fulfill, since it requires that the error network is electrically symmetrical. However, in certain practical situations, an approximation $s_{11} \approx s_{22}$ can be valid.

An additional measurement is proposed in order to drop the requirement (5.7) by producing a second test structure and thus gaining the information about the second port. The additional test structure could be generated by mirroring the error network and connecting it to the original error network. Such a structure, referred to as β, is presented in Fig. 5.3.

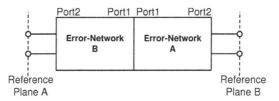

Fig. 5.3 Additional extended thru de-embedding test structure β.

Due to symmetry, ports are swapped and (5.4) is rewritten as follows

$$S_\beta = \begin{bmatrix} s_{11,\beta} & s_{12,\beta} \\ s_{21,\beta} & s_{22,\beta} \end{bmatrix} = \begin{bmatrix} s_{22} - \frac{s_{11}s_{12}s_{21}}{s_{11}^2 - 1} & -\frac{s_{12}s_{21}}{s_{11}^2 - 1} \\ -\frac{s_{12}s_{21}}{s_{11}^2 - 1} & s_{22} - \frac{s_{11}s_{12}s_{21}}{s_{11}^2 - 1} \end{bmatrix}. \tag{5.10}$$

By measuring S-parameters of the test structures in Fig. 5.2 and 5.3 one gets four equations and four variables. Equations (5.4) and (5.10) contain only the product of $s_{12}s_{21}$. Thus, the parameters s_{12} and s_{21} of the error box are inseparable and cannot be determined unambiguously. Therefore, the reciprocity assumption (5.6) still has to be applied. Referring to the measured S-parameter matrices of the two test structures as S_α and S_β, the following system of equations is obtained

$$s_{11} - \frac{s_{22}s_{12}s_{21}}{s_{22}^2 - 1} = s_{11,\alpha}$$

$$-\frac{s_{12}s_{21}}{s_{22}^2 - 1} = s_{12,\alpha}$$

$$s_{22} - \frac{s_{11}s_{12}s_{21}}{s_{11}^2 - 1} = s_{11,\beta} \qquad (5.11)$$

$$-\frac{s_{12}s_{21}}{s_{11}^2 - 1} = s_{12,\beta}.$$

Solution of (5.11) is straightforward

$$s_{11} = \frac{s_{11,\alpha} - s_{11,\beta} \cdot s_{12,\alpha}}{1 - s_{12,\alpha} \cdot s_{12,\beta}},$$

$$s_{22} = \frac{s_{11,\beta} - s_{11,\alpha} \cdot s_{12,\beta}}{1 - s_{12,\alpha} \cdot s_{12,\beta}}, \qquad (5.12)$$

$$s_{12} = s_{21} = \pm\sqrt{s_{12,\alpha}\left(1 - s_{22}^2\right)}.$$

It has to be pointed out that not in all cases it is possible to create the inverse setup as described in Fig. 5.3. However, apart from the de-embedding, this approach can be useful for simplifying measurements of asymmetrical fixtures. It can be applied e.g. for characterization of devices with different connectors on either side (in coaxial measurement setup) or with adapters of different pitch (in on-wafer measurements).

5.1.1.2 Result Verification

The above approach is confirmed by simulation of an on-chip 0.22 nH inductor, designed in Infineon's C11N technology. The layout of the inductor is presented in Fig. 5.4. The inductor is realized as a spiral rectangular coil with two turns on the top aluminium layer with a connection to the inner layer. The outer horizontal dimension of the inductor is 57 µm, the outer vertical dimension is 47 µm, the width of the trace is 6 µm and the separation between the traces is 2.6 µm.

The classical cascade-based Thru approach uses the test structure having the inductors connected back-to-back, as shown in Fig. 5.5. The inductor is treated in this case as an error box. The proposed extended Thru version is more appropriate in this case, since on-chip inductors with undercrossing are usually asymmetrical devices that do not fulfill (5.7). Therefore, an additional test structure with mirrored ports, presented in Fig. 5.6, is used.

The test structures are simulated in the Sonnet EM field solver in the frequency range 1–40 GHz. Eq. (5.9) and (5.12) are applied to de-embed the inductor S-parameters using the Thru and extended Thru approach, respectively. For comparison, shown in Fig. 5.7, the quality factors and inductances are calculated and compared with the quality factor and inductance of the directly simulated inductor in

5.1 S-parameter De-embedding Techniques 53

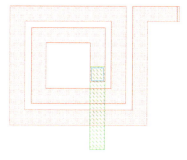

Fig. 5.4 Layout of the spiral 0.22 nH inductor in CMOS.

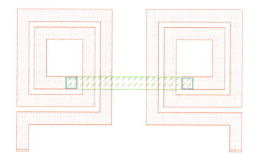

Fig. 5.5 First test structure for Thru de-embedding of coil in CMOS.

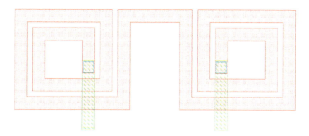

Fig. 5.6 Second test structure for Thru de-embedding of coil in CMOS.

Fig. 5.4. As can be seen, the additional information obtained from the second structure in Fig. 5.6 allows more accurate results for the extracted network to be obtained.

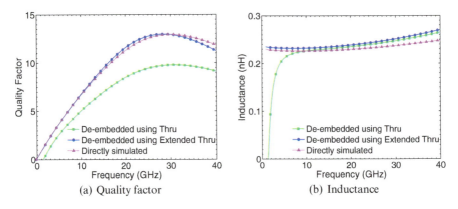

Fig. 5.7 Comparison of results de-embedded using Thru and extended Thru.

5.1.2 De-embedding of Differential Devices using cascade-based Two-Port Techniques

Accurate on-wafer measurement of differential circuits at microwave frequencies is a challenge. In order to characterize differential devices, the mixed-mode S-parameters theory has been formulated [10] and measurement techniques using a pure-mode VNA have been developed [11]. Additionally, the multimode TRL calibration technique has been devised for characterization of multiport devices by means of multimode networks [12]. However, only few works describe the de-embedding of differential devices from four-port S-parameter measurements [13]. This work proposes applying classical cascade-based two-port techniques to four-port S-parameters for de-embedding of differential networks. Under certain conditions, presented in [14] and described in detail in this section, it is possible to separate the modes and consider only the differential S-parameters, thus reducing the measured 4×4 matrix to a 2×2 matrix and considering the error box as a two-port network.

5.1.2.1 Theory

Similarly to section 5.1.1, the expected S-parameters of a four-port VNA measurement can be considered as a chain connection of networks shown in Fig. 5.8. The four-port error networks A and B have to be de-embedded. It is assumed that the error network B represents the mirrored version of the network A and that both networks are uncoupled. The reference planes A and B are defined by a standard VNA calibration technique and the reference planes for the device under test (DUT) in Fig. 5.8 are set by a de-embedding technique.

The nodal S-parameter matrix S_n of a general four-port network, defined as

5.1 S-parameter De-embedding Techniques

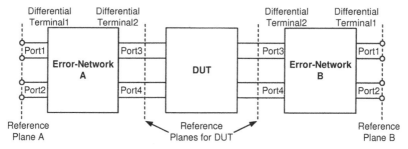

Fig. 5.8 Chain connection of four-port DUT and error networks.

$$S_n = \begin{bmatrix} s_{11} & s_{12} & s_{13} & s_{14} \\ s_{21} & s_{22} & s_{23} & s_{24} \\ s_{31} & s_{32} & s_{33} & s_{34} \\ s_{41} & s_{42} & s_{43} & s_{44} \end{bmatrix}, \quad (5.13)$$

can be easily converted into the modal form

$$S_m = \begin{bmatrix} s_{11}^{dd} & s_{11}^{dc} & s_{12}^{dd} & s_{12}^{dc} \\ s_{11}^{cd} & s_{11}^{cc} & s_{12}^{cd} & s_{12}^{cc} \\ s_{21}^{dd} & s_{21}^{dc} & s_{22}^{dd} & s_{22}^{dc} \\ s_{21}^{cd} & s_{21}^{cc} & s_{22}^{cd} & s_{22}^{cc} \end{bmatrix}, \quad (5.14)$$

where indices 1 and 2 describe the differential terminals, each containing two single-ended ports, as presented in Fig. 5.8, using the following transformation

$$S_m = \begin{bmatrix} M & E \\ E & M \end{bmatrix}^{-1} \cdot S_n \cdot \begin{bmatrix} M & E \\ E & M \end{bmatrix}, \quad (5.15)$$

where E is an empty 2×2 sub-matrix and M is defined as follows

$$M = \frac{1}{\sqrt{2}} \begin{bmatrix} -1 & 1 \\ 1 & 1 \end{bmatrix}. \quad (5.16)$$

This definition is equivalent to the classical mixed-mode S-parameter conversion presented in [10], but in this case the order of the wave vectors has been modified for convenience of cascading. The terms s_{dd} and s_{cc} describe differential and common-mode S-parameters respectively, whilst s_{dc} and s_{cd} describe the mode conversion. Mode conversion can occur due to asymmetry within the differential interconnect or due to unbalanced loading [15]. If a differential structure is fully symmetrical for the signal propagation, the modes are decoupled and for a corresponding excitation pure differential mode or pure common-mode can propagate. If the structure is asymmetrical, e.g. if one transmission line in a differential pair is longer than the other one or they have differing widths, part of the energy launched by differential excitation is lost due to the conversion into the common-mode.

In practice the differential transitions used as the error boxes are usually designed to be symmetrical with respect to the z-axis (in the direction of propagation). The asymmetry that might stem from manufacturing tolerances or from the signal path is usually negligible for frequencies up to several tens of gigahertz. As an example of a symmetrical differential error box, an on-chip transition conducting the differential signals to the active circuit area is presented in Fig. 5.9(a). An additional example of a symmetrical transition is a test board for the characterization of a differential LNA using coaxial signaling, as presented in Fig. 5.9(b).

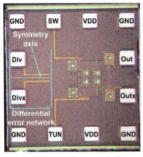

(a) Active differential chip

(b) Differential test board

Fig. 5.9 Differential error networks with small mode conversion.

One can apply these considerations and refer to S_m as the modal S-parameters of the error network A. Therefore, under the assumption of a symmetrical differential error box, the S-parameter terms describing mode conversion shall be considerably smaller than the terms corresponding to mode propagation. Combining this with reciprocity of the error box, this condition can be formulated as follows

$$\begin{aligned} s_{11}^{dc} \approx s_{11}^{cd} &\approx 0, \\ s_{12}^{dc} \approx s_{12}^{cd} &\approx 0, \\ s_{21}^{dc} \approx s_{21}^{cd} &\approx 0, \\ s_{22}^{dc} \approx s_{22}^{cd} &\approx 0. \end{aligned} \qquad (5.17)$$

Thus, the modal matrix (5.14) can be approximated and simplified to the following form

$$\begin{bmatrix} b_1^{dm} \\ b_1^{cm} \\ b_2^{dm} \\ b_2^{cm} \end{bmatrix} = S_m \begin{bmatrix} a_1^{dm} \\ a_1^{cm} \\ a_2^{dm} \\ a_2^{cm} \end{bmatrix} = \begin{bmatrix} S_{11} & S_{12} \\ S_{21} & S_{22} \end{bmatrix} \begin{bmatrix} a_1^{dm} \\ a_1^{cm} \\ a_2^{dm} \\ a_2^{cm} \end{bmatrix} \approx \begin{bmatrix} s_{11}^{dd} & 0 & s_{12}^{dd} & 0 \\ 0 & s_{11}^{cc} & 0 & s_{12}^{cc} \\ s_{21}^{dd} & 0 & s_{22}^{dd} & 0 \\ 0 & s_{21}^{cc} & 0 & s_{22}^{cc} \end{bmatrix} \begin{bmatrix} a_1^{dm} \\ a_1^{cm} \\ a_2^{dm} \\ a_2^{cm} \end{bmatrix}. \qquad (5.18)$$

5.1 S-parameter De-embedding Techniques

As can be observed, the sub-matrices in (5.18) are diagonal. Thus, the matrix S_m preserves its form upon conversion into T-parameters, given by [16]

$$T = \begin{bmatrix} S_{12} - S_{11}S_{21}^{-1}S_{22} & S_{11}S_{21}^{-1} \\ -S_{21}^{-1}S_{22} & S_{21}^{-1} \end{bmatrix}. \quad (5.19)$$

Therefore, the differential parameters remain separated from common-mode parameters. Taking into consideration that when the network is mirrored, ports 2 and 1 are swapped and assuming reciprocity of the error networks $s_{12}^{dd} = s_{21}^{dd}$ and $s_{12}^{cc} = s_{21}^{cc}$, the S-parameters of A and B can now be easily converted to the T-parameters as

$$T_A = \begin{bmatrix} t_{11}^{dd} & 0 & t_{12}^{dd} & 0 \\ 0 & t_{11}^{cc} & 0 & t_{12}^{cc} \\ t_{21}^{dd} & 0 & t_{22}^{dd} & 0 \\ 0 & t_{21}^{cc} & 0 & t_{22}^{cc} \end{bmatrix} = \begin{bmatrix} s_{12}^{dd} - \frac{s_{11}^{dd} s_{22}^{dd}}{s_{12}^{dd}} & 0 & \frac{s_{11}^{dd}}{s_{12}^{dd}} & 0 \\ 0 & s_{12}^{cc} - \frac{s_{11}^{cc} s_{22}^{cc}}{s_{12}^{cc}} & 0 & \frac{s_{11}^{cc}}{s_{12}^{cc}} \\ -\frac{s_{22}^{dd}}{s_{12}^{dd}} & 0 & \frac{1}{s_{12}^{dd}} & 0 \\ 0 & -\frac{s_{22}^{cc}}{s_{12}^{cc}} & 0 & \frac{1}{s_{12}^{cc}} \end{bmatrix}, \quad (5.20)$$

$$T_B = \begin{bmatrix} t_{11}^{dd} & 0 & -t_{21}^{dd} & 0 \\ 0 & t_{11}^{cc} & 0 & -t_{21}^{cc} \\ -t_{12}^{dd} & 0 & t_{22}^{dd} & 0 \\ 0 & -t_{12}^{cc} & 0 & t_{22}^{cc} \end{bmatrix} = \begin{bmatrix} s_{12}^{dd} - \frac{s_{22}^{dd} s_{11}^{dd}}{s_{12}^{dd}} & 0 & \frac{s_{22}^{dd}}{s_{12}^{dd}} & 0 \\ 0 & s_{12}^{cc} - \frac{s_{22}^{cc} s_{11}^{cc}}{s_{12}^{cc}} & 0 & \frac{s_{22}^{cc}}{s_{12}^{cc}} \\ -\frac{s_{11}^{dd}}{s_{12}^{dd}} & 0 & \frac{1}{s_{12}^{dd}} & 0 \\ 0 & -\frac{s_{11}^{cc}}{s_{12}^{cc}} & 0 & \frac{1}{s_{12}^{cc}} \end{bmatrix}. \quad (5.21)$$

Now the four-port error networks A and B are cascaded in order to construct the thru standard, as shown in Fig. 5.10.

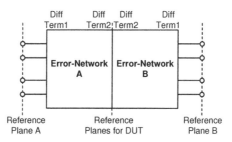

Fig. 5.10 Differential thru standard.

The T-matrix of the thru standard is simply given by multiplication of (5.20) and (5.21) and can be written in a simplified form as follows

$$T_{\text{Thru}} = T_{\text{A}} \cdot T_{\text{B}} = \begin{bmatrix} (t_{11}^{dd})^2 - (t_{12}^{dd})^2 & 0 & t_{12}^{dd}t_{22}^{dd} - t_{11}^{dd}t_{21}^{dd} & 0 \\ 0 & (t_{11}^{cc})^2 - (t_{12}^{cc})^2 & 0 & t_{12}^{cc}t_{22}^{cc} - t_{11}^{cc}t_{21}^{cc} \\ t_{11}^{dd}t_{21}^{dd} - t_{12}^{dd}t_{22}^{dd} & 0 & (t_{22}^{dd})^2 - (t_{21}^{dd})^2 & 0 \\ 0 & t_{11}^{cc}t_{21}^{cc} - t_{12}^{cc}t_{22}^{cc} & 0 & (t_{22}^{cc})^2 - (t_{21}^{cc})^2 \end{bmatrix}. \quad (5.22)$$

As can be observed, this results again in a matrix containing four diagonal submatrices. Thus, the differential and common-mode parameters remain separated and the operations can be also reduced to two-port matrices containing either mode. Obviously, upon conversion back to S-parameters the matrix maintains the same form.

Similar considerations are valid for the S-parameters of the line standard, presented in Fig. 5.11. The T-matrices of the error networks A and B have the

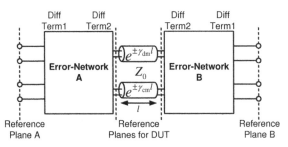

Fig. 5.11 Differential line standard.

form (5.20) and (5.21), respectively, and the transmission matrix of the line is given by a diagonal matrix containing the mode propagation coefficients $e^{\pm \gamma_{dm} l}$ and $e^{\pm \gamma_{cm} l}$. This would simply result in multiplication of the entries in (5.22) by the corresponding propagation coefficients.

Therefore, under condition of negligible mode conversion on the de-embedded error networks, it is valid to treat the modes separately and apply the classical two-port TL or TRL procedures to the S-parameters of each mode separately. Obviously, the presented considerations can be further expanded to similar techniques, such as e.g. multiline TRL [17]. The additional lines follow the same reasoning and under the assumption of a weak mode conversion the S-parameters corresponding to different modes remain separated.

In practice, it is difficult to obtain the four-port S-parameters of an error-network in order to verify, whether the conditions in (5.17) are fulfilled. Since the terms corresponding to the mode conversion in (5.22) remain negligible, if assumption (5.17) is fulfilled, one can formulate an equivalent condition on the applicability of the mode separation approach using the measured four-port S-parameters of the thru or line structures. The differential on-chip error boxes are usually designed to be symmetrical and the measured mode conversion terms commonly remain below

5.1 S-parameter De-embedding Techniques

$-30\,\text{dB}$. Therefore, a practical condition equivalent to (5.17) can be defined by observing the mode conversion S-parameters of the thru or line standards, obtained by conversion of the measured four-port nodal matrix into the mixed-mode form. The parameters s_{11}^{dc}, s_{11}^{cd}, s_{12}^{dc}, s_{12}^{cd}, s_{21}^{dc}, s_{21}^{cd}, s_{22}^{dc}, s_{22}^{cd} should be negligible over the whole frequency range. Usually, it is sufficient that these parameters are less than $-30\,\text{dB}$ for the presented considerations to be applicable. These conditions are often fulfilled for on-chip de-embedding structures. However, if these conditions are not fulfilled, the multimode TRL [12] has to be applied for de-embedding of differential S-parameters.

In the next step DUT T-parameters are de-embedded using

$$\tilde{T}_{\text{DUT}} = \tilde{T}_A^{\ -1} \cdot T_{\text{meas}} \cdot \tilde{T}_B^{\ -1}, \tag{5.23}$$

where $\tilde{T}_A$ and $\tilde{T}_B$ are the T-parameters of the error networks A and B, estimated using any two-port de-embedding technique, and T_{meas} are the measured T-parameters of the chain connection in Fig. 5.8 before de-embedding.

Now two separate cases have to be considered: of a DUT having negligible mode conversion and of a DUT having non-negligible mode conversion.

When a DUT has negligible mode conversion, as expected from a differential amplifier or a symmetrical passive structure, its T-matrix can be written as

$$T_{\text{DUT}} \approx \begin{bmatrix} t_{11,\text{DUT}}^{dd} & 0 & t_{12,\text{DUT}}^{dd} & 0 \\ 0 & t_{11,\text{DUT}}^{cc} & 0 & t_{12,\text{DUT}}^{cc} \\ t_{21,\text{DUT}}^{dd} & 0 & t_{22,\text{DUT}}^{dd} & 0 \\ 0 & t_{21,\text{DUT}}^{cc} & 0 & t_{22,\text{DUT}}^{cc} \end{bmatrix}. \tag{5.24}$$

Therefore, under assumption that T-matrices describing the error networks A and B have the same form as T_{DUT} in (5.24), the differential and common-mode parameters of the measured chain connection in Fig. 5.8 remain separated and matrix T_{meas} in (5.23) maintains the same form as T_{DUT}. Thus, in case that differential parameters of the DUT are of interest, it is sufficient to consider only the differential parameters and the matrices in (5.23) become 2×2

$$\tilde{T}_{\text{DUT}}^{dd} = \begin{bmatrix} t_{11}^{dd} & t_{12}^{dd} \\ t_{21}^{dd} & t_{22}^{dd} \end{bmatrix}^{-1} \begin{bmatrix} t_{11,\text{meas}}^{dd} & t_{12,\text{meas}}^{dd} \\ t_{21,\text{meas}}^{dd} & t_{22,\text{meas}}^{dd} \end{bmatrix} \begin{bmatrix} t_{11}^{dd} & -t_{21}^{dd} \\ -t_{12}^{dd} & t_{22}^{dd} \end{bmatrix}^{-1}, \tag{5.25}$$

where t_{11}^{dd}, t_{12}^{dd}, t_{21}^{dd}, t_{22}^{dd} are the differential T-parameters of the error boxes that can be obtained using any two-port cascade-based technique and $t_{11,\text{meas}}^{dd}, t_{12,\text{meas}}^{dd}$, $t_{21,\text{meas}}^{dd}, t_{22,\text{meas}}^{dd}$ are the differential T-parameters of the setup described in Fig. 5.8, measured using a calibrated VNA.

However, if mode conversion of a DUT is not negligible, its T-matrix has to be considered as a full 4×4 matrix

$$
T_{\text{DUT}} \approx
\begin{bmatrix}
t^{dd}_{11,\text{DUT}} & t^{dc}_{11,\text{DUT}} & t^{dd}_{12,\text{DUT}} & t^{dc}_{12,\text{DUT}} \\
t^{cd}_{11,\text{DUT}} & t^{cc}_{11,\text{DUT}} & t^{cd}_{12,\text{DUT}} & t^{cc}_{12,\text{DUT}} \\
t^{dd}_{21,\text{DUT}} & t^{dc}_{21,\text{DUT}} & t^{dd}_{22,\text{DUT}} & t^{dc}_{22,\text{DUT}} \\
t^{cd}_{21,\text{DUT}} & t^{cc}_{21,\text{DUT}} & t^{cd}_{22,\text{DUT}} & t^{cc}_{22,\text{DUT}}
\end{bmatrix}.
\tag{5.26}
$$

Therefore, the differential and common-mode parameters of the measured chain connection in Fig. 5.8 get mixed and cannot be separated and matrix T_{meas} has to be treated as a full 4×4 matrix. The matrices in (5.23) can be thus written as

$$
\tilde{T}_{\text{DUT}} =
\begin{bmatrix}
t^{dd}_{11} & 0 & t^{dd}_{12} & 0 \\
0 & t^{cc}_{11} & 0 & t^{cc}_{12} \\
t^{dd}_{21} & 0 & t^{dd}_{22} & 0 \\
0 & t^{cc}_{21} & 0 & t^{cc}_{22}
\end{bmatrix}^{-1}
\begin{bmatrix}
t^{dd}_{11,\text{meas}} & t^{dc}_{11,\text{meas}} & t^{dd}_{12,\text{meas}} & t^{dc}_{12,\text{meas}} \\
t^{cd}_{11,\text{meas}} & t^{cc}_{11,\text{meas}} & t^{cd}_{12,\text{meas}} & t^{cc}_{12,\text{meas}} \\
t^{dd}_{21,\text{meas}} & t^{dc}_{21,\text{meas}} & t^{dd}_{22,\text{meas}} & t^{dc}_{22,\text{meas}} \\
t^{cd}_{21,\text{meas}} & t^{cc}_{21,\text{meas}} & t^{cd}_{22,\text{meas}} & t^{cc}_{22,\text{meas}}
\end{bmatrix}
\begin{bmatrix}
t^{dd}_{11} & 0 & -t^{dd}_{21} & 0 \\
0 & t^{cc}_{11} & 0 & -t^{cc}_{21} \\
-t^{dd}_{12} & 0 & t^{dd}_{22} & 0 \\
0 & -t^{cc}_{12} & 0 & t^{cc}_{22}
\end{bmatrix}^{-1},
\tag{5.27}
$$

where the T-matrix, describing the error network A, is obtained by applying any cascade-based de-embedding technique twice: once to differential and once to common-mode parameters and combining the 2×2 matrices into a 4×4 matrix.

5.1.2.2 Result Verification

The above considerations are confirmed in measurement and simulation. Firstly, a 2:1 transformer, fabricated in C11N process, exemplifies a DUT with negligible mode conversion. Secondly, an asymmetrical on-chip differential line, simulated in the same process, describes a case of a DUT with non-negligible mode conversion.

In the first example, on-wafer test structures have been produced in Infineon's C11N process. They include short, open, transmission line and thru. The measurements have been performed on-wafer using Cascade Microtech Infinity probes with $100\,\mu\text{m}$ pitch in GSSG configuration and Agilent's four-port VNA up to $50\,\text{GHz}$, calibrated using the four-port SOLT technique.

The DUT in this case is a 2:1 transformer, presented in Fig. 5.12. The chip is also manufactured in the C11N process. The primary ports, P+ and P-, are located on the left hand side. The secondary ports, S+ and S-, are located on the right hand side. The outer diameter is $92\,\mu\text{m}$ and the inner diameter is $50\,\mu\text{m}$. The lateral spacing between the turns is $2.5\,\mu\text{m}$. The conductor-width of the primary windings is $6\,\mu\text{m}$ and of the secondary winding is $4\,\mu\text{m}$.

The cascade-based TRL [8], simplified TL [18] and the lumped-element based two-step Short-Open [2] techniques are applied for comparison. In order to be able to perform a comparison with the aforementioned lumped-element technique, the port impedances of the S-parameters, de-embedded using TL and TRL, have been re-normalized to $100\,\Omega$.

5.1 S-parameter De-embedding Techniques

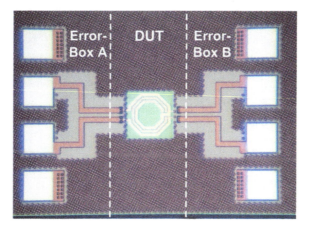

Fig. 5.12 Chip micrograph of the 2:1 transformer (553 μm × 430 μm).

The equivalent inductance of the primary and secondary windings, de-embedded using various techniques, is presented in Fig. 5.13 and Fig. 5.14, respectively.

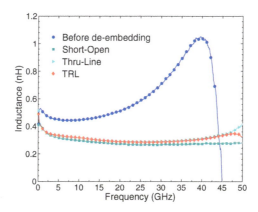

Fig. 5.13 Primary side inductance of the 2:1 transformer in CMOS.

As can be observed, the comparison shows a very good match over a wide range of frequencies. However, the difference between the lumped-element based Short-Open and cascade-based TL and TRL methods is smaller than expected. This is due to the fact that the physical size of the error box is much smaller compared to the wavelength and thus Short-Open is still sufficiently accurate even at 50 GHz.

The larger discrepancy for the secondary side stems from the inaccuracy of TL and TRL methods using a single line standard at lower frequencies. The accuracy of TL and TRL can be improved by using lines with different lengths [17].

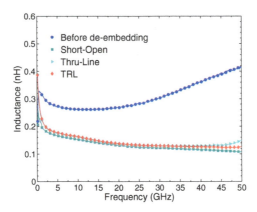

Fig. 5.14 Secondary side inductance of the 2:1 transformer in CMOS.

The mode conversion parameters of the thru and line standards were measured to be below −35 dB over the whole frequency range. Thus, the condition (5.17) was fulfilled and the presented mode separation considerations were applicable.

In the second example, an asymmetrical differential on-chip microstrip line has been simulated using a full-wave Ansoft HFSS field-solver. The line has been realized in the top layer of the C11N layer stack-up. The differential impedance of the line has been designed to be close to $100\,\Omega$. The separation between the traces and their widths are 10 μm. The length of one line of the differential pair is 1220 μm, whilst the length of the other line is 2076 μm. Mode conversion of the DUT grows with increasing frequency. Its simulated value exceeds −35 dB at 1 GHz and reaches −5 dB at 50 GHz. The DUT including the symmetrical differential error boxes is shown in Fig. 5.15.

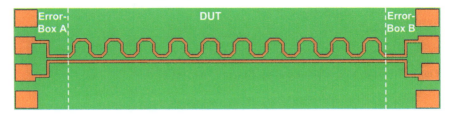

Fig. 5.15 Simulated asymmetrical differential DUT with error boxes.

The TRL technique has been applied twice to the simulated S-parameters of thru, line and open standards in order to obtain 2×2 matrices of the differential and common-mode error box S-parameters. The impact of the error boxes has been removed using (5.27) and the de-embedded results have been compared with directly simulated S-parameters of the DUT. Fig. 5.16 presents the comparison of the differential transmission S-parameter of the asymmetrical line. As can be seen, the proposed approach of de-embedding four-port DUTs using two-port methods shows accurate results also for a DUT with non-negligible mode conversion.

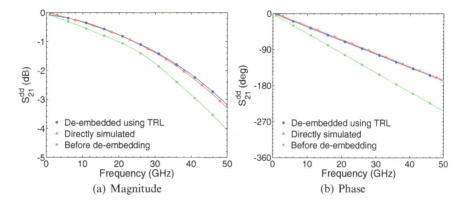

Fig. 5.16 Comparison of the differential transmission parameter.

5.2 Differential Measurements using Baluns

The characterization of active differential devices requires costly four-port or true-differential network analyzers. Thus, a common method is to measure the S-parameters of two ports, while the other two ports are terminated, as described in [19]. This method has the disadvantages of complexity, redundancy in the interpretation of measurement results and time consumption. It requires performing a sequence of six measurements to obtain the sixteen single-ended S-parameters and to convert them into the modal representation.

Another very common approach is to introduce baluns to convert between single-ended and differential signals and perform the measurement using a lower cost two-port VNA or a spectrum analyzer [20]. This approach is popular due to its simplicity, but has the disadvantage of introducing additional components that cannot be easily de-embedded. Back-to-back connection of baluns is usually implemented in order to account for their impact using the widely used *Insertion Loss* de-embedding method. This approach is straightforward, but has a moderate accuracy and provides only the amplitude information.

This section presents a generalized analytical analysis of the back-to-back baluns interconnection measurement setup and applies the results for the error estimation of the *Insertion Loss* technique. Additionally, it shows analytically and in measurement that the frequency characteristics of a differential circuit measured using baluns, might be distorted due to numerous contributions that cannot be easily accounted for.

5.2.1 Theoretical Analysis

The measurement setup of a differential device using a calibrated two-port VNA or a spectrum analyzer can be considered as a chain connection of networks, as shown in Fig. 5.17.

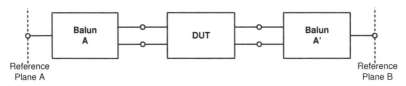

Fig. 5.17 Two-port measurement setup of a differential device using baluns.

Similarly to the two-port and four-port error networks in Fig. 5.1 and Fig. 5.8, respectively, the baluns A and A' are considered to be three-port error networks. These networks have to be characterized and their impact has to be removed from the measured results of the DUT in Fig. 5.17.

For simplicity of the analysis it has to be assumed that the balun A' is the mirrored version of the balun A, and that the networks are uncoupled. Furthermore, for easier interpretation of the results the modal balun S-parameters are considered. Using these assumptions one can derive the expected S-parameters of the *back-to-back* interconnection in Fig. 5.18. This setup is required by the widely used *Insertion Loss* technique in order to evaluate the balun characteristics.

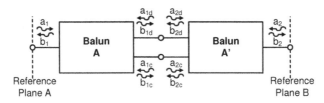

Fig. 5.18 Back-to-back balun interconnection setup.

The *Insertion Loss* technique can be implemented using a two-port VNA or by means of a spectrum analyzer. In this technique the balun's insertion loss in decibel IL_{bln} (dB) is obtained by measuring the total insertion loss of the back-to-back interconnection in Fig. 5.18 in decibel IL_{b2b} (dB) and dividing it by two [20]

5.2 Differential Measurements using Baluns

$$\text{IL}_{\text{bln}}\,(\text{dB}) \approx \frac{\text{IL}_{\text{b2b}}\,(\text{dB})}{2}. \tag{5.28}$$

For a particular case when the DUT is a differential amplifier, its de-embedded differential gain $\tilde{G}_{\text{DUT}}$ is given using the *Insertion Loss* method by

$$\tilde{G}_{\text{DUT}}\,(\text{dB}) \approx G_{\text{meas}}\,(\text{dB}) - \text{IL}_{\text{b2b}}\,(\text{dB}), \tag{5.29}$$

where $G_{\text{meas}}\,(\text{dB})$ is the measured differential gain in decibel of the setup in Fig. 5.17. Obviously, this expression is only valid under very restricting assumptions and when all the components are matched to $50\,\Omega$.

Thus, under the assumption of a good port matching, high reverse isolation and negligible mode conversion, required by the *Insertion Loss* method, Eq. (5.29) can be related to the S-parameters measured in a $50\,\Omega$ environment and rewritten as follows

$$\left|\tilde{s}_{21,\text{dd}}^{\text{DUT}}\right|^2 \approx \frac{|s_{21,\text{meas}}|^2}{|s_{21,\text{b2b}}|^2}, \tag{5.30}$$

where $\tilde{s}_{21,\text{dd}}^{\text{DUT}}$ is the de-embedded gain of the amplifier, $s_{21,\text{meas}}$ is the transmission S-parameter measured for the setup in Fig. 5.17 and $s_{21,\text{b2b}}$ is the transmission S-parameter measured for the back-to-back setup in Fig. 5.18.

The following analysis provides analytical expressions for the S-parameters expected to be measured for the back-to-back connection of the baluns and for the DUT measurement with baluns. The derived expressions are used in (5.30) and the error of gain de-embedding using the *Insertion Loss* approach is evaluated.

5.2.1.1 Back-to-Back Measurement

Independent of realization, baluns are considered to be three-port devices. Nodal to modal S-parameter conversion for three-port devices is given in [21]. The internal ports of the multiport setup describe the differential and common-mode wave amplitudes. Considering reciprocity of passive baluns exists $s_{1d} = s_{d1}$, $s_{1c} = s_{c1}$ and $s_{dc} = s_{cd}$. Therefore, the S-parameters of a balun can be written as follows

$$S_{\text{bln}} = \begin{bmatrix} s_{11} & s_{1d} & s_{1c} \\ s_{1d} & s_{dd} & s_{dc} \\ s_{1c} & s_{dc} & s_{cc} \end{bmatrix}, \tag{5.31}$$

where index 1 denotes the single-ended port, index d denotes the differential port and index c denotes the common-mode port.

Following the Multiport Connection Method for evaluating circuit parameters of an arbitrarily interconnected network [22], [23], the rows and columns of the overall S-parameter matrix S_{b2b} are ordered so that the wave variables are separated into groups corresponding to the e external ports and i internally connected ports. Thus, the wave relation of the multiport network can be written as

$$\begin{bmatrix} b_e \\ b_i \end{bmatrix} = \begin{bmatrix} S_{ee} & S_{ei} \\ S_{ie} & S_{ii} \end{bmatrix} \begin{bmatrix} a_e \\ a_i \end{bmatrix},$$ (5.32)

where b_e and a_e are the waves at the e external ports and b_i and a_i are the waves at the i internal ports. For the setup in Fig. 5.18 the relation can be written as follows

$$\begin{bmatrix} b_1 \\ b_2 \\ b_{1d} \\ b_{1c} \\ b_{2c} \\ b_{2d} \end{bmatrix} = \begin{bmatrix} s_{11} & 0 & s_{1d} & s_{1c} & 0 & 0 \\ 0 & s_{11} & 0 & 0 & s_{1c} & s_{1d} \\ s_{1d} & 0 & s_{dd} & s_{dc} & 0 & 0 \\ s_{1c} & 0 & s_{dc} & s_{cc} & 0 & 0 \\ 0 & s_{1c} & 0 & 0 & s_{cc} & s_{dc} \\ 0 & s_{1d} & 0 & 0 & s_{dc} & s_{dd} \end{bmatrix} \begin{bmatrix} a_1 \\ a_2 \\ a_{1d} \\ a_{1c} \\ a_{2c} \\ a_{2d} \end{bmatrix},$$ (5.33)

where a_1, a_2 and b_1, b_2 are the incident and reflected wave variables at the external ports, $a_{1d}, a_{1c}, a_{2d}, a_{2c}$ and $b_{1d}, b_{1c}, b_{2d}, b_{2c}$ are the incident and reflected wave variables at the internal ports of the multiport, as depicted in Fig. 5.18, and $s_{11}, s_{1d}, s_{1c}, s_{dd}, s_{dc}, s_{cc}$ are the S-parameters of the balun defined in (5.31). The interconnection of the internal ports is described by the matrix Γ, defined as $b_i = \Gamma a_i$. For the setup in Fig. 5.18, Γ can be explicitly written as follows

$$\begin{bmatrix} b_{1d} \\ b_{1c} \\ b_{2c} \\ b_{2d} \end{bmatrix} = \begin{bmatrix} 0 & 0 & 0 & 1 \\ 0 & 0 & 1 & 0 \\ 0 & 1 & 0 & 0 \\ 1 & 0 & 0 & 0 \end{bmatrix} \begin{bmatrix} a_{1d} \\ a_{1c} \\ a_{2c} \\ a_{2d} \end{bmatrix}.$$ (5.34)

The scattering matrix of the back-to-back setup is thus given by

$$S_{b2b} = \begin{bmatrix} s_{11,b2b} & s_{12,b2b} \\ s_{21,b2b} & s_{11,b2b} \end{bmatrix} = S_{ee} + S_{ei} (\Gamma - S_{ii})^{-1} S_{ie}.$$ (5.35)

Substituting the matrices from (5.33) and (5.34) into (5.35), and after some mathematical manipulation one arrives at the expressions (5.36) and (5.37) for transmission and reflection S-parameters, respectively.

$$s_{12,b2b} = s_{21,b2b} = \frac{\left(1 - s_{dc}^2 - s_{cc}^2\right) s_{1d}^2 + \left(1 - s_{dd}^2 - s_{dc}^2\right) s_{1c}^2 + 2 s_{1c} s_{dc} s_{1d} \left(s_{cc} + s_{dd}\right)}{\left[(s_{dd} + 1)(s_{cc} + 1) - s_{dc}^2\right] \left[(s_{cc} - 1)(s_{dd} - 1) - s_{dc}^2\right]}$$ (5.36)

$$s_{11,b2b} = s_{22,b2b} = \frac{s_{11} \left[\left(s_{cc}^2 - 1\right)\left(s_{dd}^2 - 1\right) + \left(s_{dc}^2 - 2 s_{dd} s_{cc} - 2\right) s_{dc}^2\right]}{\left[(s_{dd} + 1)(s_{cc} + 1) - s_{dc}^2\right] \left[(s_{cc} - 1)(s_{dd} - 1) - s_{dc}^2\right]}$$
$$+ \frac{s_{1c}^2 \left[s_{dd} s_{dc}^2 + s_{cc} \left(1 - s_{dd}^2\right)\right] + s_{1d}^2 \left[s_{cc} s_{dc}^2 + s_{dd} \left(1 - s_{cc}^2\right)\right] + 2 s_{1d} s_{1c} s_{dc} \left(s_{cc} s_{dd} - s_{dc}^2 + 1\right)}{\left[(s_{dd} + 1)(s_{cc} + 1) - s_{dc}^2\right] \left[(s_{cc} - 1)(s_{dd} - 1) - s_{dc}^2\right]}$$ (5.37)

Due to the obvious complexity of the expressions it is not possible to isolate the S-parameters of a single balun without further simplification. This leads naturally to de-embedding inaccuracy.

5.2.1.2 DUT Measurement

Now the inherent inaccuracy of DUT de-embedding due to implementation of baluns is analyzed. The DUT in Fig. 5.17 is considered to be a differential amplifier. The Multiport Connection Method is applied again to obtain the expected measured S-parameters of the setup in Fig. 5.17

$$S_{\text{meas}} = \begin{bmatrix} s_{11,\text{meas}} & s_{12,\text{meas}} \\ s_{21,\text{meas}} & s_{11,\text{meas}} \end{bmatrix} = S_{ee} + S_{ei} \left(\Gamma - S_{ii} \right)^{-1} S_{ie}. \tag{5.38}$$

For simplicity it is assumed that the DUT has negligible mode conversion, which is usually desired for a differential amplifier or a symmetrical passive structure. Thus, the sub-matrices of the multiport matrix S_{ee}, S_{ei}, S_{ie} and S_{ii} are given by

$$\begin{bmatrix} S_{ee} & S_{ei} \\ S_{ie} & S_{ii} \end{bmatrix} = \begin{bmatrix} s_{11} & 0 & 0 & s_{1c} & 0 & s_{1d} & 0 & 0 & 0 & 0 \\ 0 & s_{11} & s_{1c} & 0 & s_{1d} & 0 & 0 & 0 & 0 & 0 \\ 0 & s_{1c} & s_{cc} & 0 & s_{dc} & 0 & 0 & 0 & 0 & 0 \\ s_{1c} & 0 & 0 & s_{cc} & 0 & s_{dc} & 0 & 0 & 0 & 0 \\ 0 & s_{1d} & s_{dc} & 0 & s_{dd} & 0 & 0 & 0 & 0 & 0 \\ s_{1d} & 0 & 0 & s_{dc} & 0 & s_{dd} & 0 & 0 & 0 & 0 \\ 0 & 0 & 0 & 0 & 0 & 0 & s_{11,dd}^{DUT} & s_{12,dd}^{DUT} & 0 & 0 \\ 0 & 0 & 0 & 0 & 0 & 0 & s_{21,dd}^{DUT} & s_{22,dd}^{DUT} & 0 & 0 \\ 0 & 0 & 0 & 0 & 0 & 0 & 0 & 0 & s_{11,cc}^{DUT} & s_{12,cc}^{DUT} \\ 0 & 0 & 0 & 0 & 0 & 0 & 0 & 0 & s_{21,cc}^{DUT} & s_{22,cc}^{DUT} \end{bmatrix} \tag{5.39}$$

and the matrix Γ, describing the internal ports interconnection, is given by

$$\Gamma = \begin{bmatrix} 0 & 0 & 0 & 0 & 0 & 0 & 0 & 1 \\ 0 & 0 & 0 & 0 & 0 & 0 & 1 & 0 \\ 0 & 0 & 0 & 0 & 0 & 1 & 0 & 0 \\ 0 & 0 & 0 & 0 & 1 & 0 & 0 & 0 \\ 0 & 0 & 0 & 1 & 0 & 0 & 0 & 0 \\ 0 & 0 & 1 & 0 & 0 & 0 & 0 & 0 \\ 0 & 1 & 0 & 0 & 0 & 0 & 0 & 0 \\ 1 & 0 & 0 & 0 & 0 & 0 & 0 & 0 \end{bmatrix}. \tag{5.40}$$

The S-parameters of the DUT with index cc describe common-mode and dd differential mode reflection and transmission coefficients. The measured S-parameters of the DUT with the error boxes S_{meas} are given by substituting (5.39) and (5.40) into (5.38). The resulting expressions are very lengthy, but can be considerably simplified by assuming that at the frequencies of interest the mode conversion of the balun is negligible $s_{dc} = 0$. The de-embedded forward transmission parameter, corresponding to the differential amplifier, is then given by

$$s_{21,\text{meas}} \approx s_{21,\text{dd}}^{\text{DUT}} \frac{s_{1d}^2}{1 - \left(s_{11,\text{dd}}^{\text{DUT}} + s_{22,\text{dd}}^{\text{DUT}}\right) s_{\text{dd}} + \left(s_{11,\text{dd}}^{\text{DUT}} s_{22,\text{dd}}^{\text{DUT}} - s_{12,\text{dd}}^{\text{DUT}} s_{21,\text{dd}}^{\text{DUT}}\right) s_{\text{dd}}^2}$$
$$+ s_{21,\text{cc}}^{\text{DUT}} \frac{s_{1c}^2}{1 - \left(s_{11,\text{cc}}^{\text{DUT}} + s_{22,\text{cc}}^{\text{DUT}}\right) s_{\text{cc}} + \left(s_{11,\text{cc}}^{\text{DUT}} s_{22,\text{cc}}^{\text{DUT}} - s_{12,\text{cc}}^{\text{DUT}} s_{21,\text{cc}}^{\text{DUT}}\right) s_{\text{cc}}^2}. \tag{5.41}$$

It can be easily shown that the first part of the expression corresponds to the differential s_{21}, expected to be measured if baluns are considered to be purely differential two-port networks. However, the second part of the expression shows that the common-mode gain of the amplifier is transferred through the common-mode transmission parameter of the balun and distorts the de-embedded gain. Therefore, the common-mode properties of the differential DUT also affect the de-embedding accuracy when measured using a two-port VNA and baluns.

5.2.1.3 Insertion Loss De-embedding Error

The expressions for $s_{21,\text{b2b}}$ and $s_{21,\text{meas}}$, given in (5.36) and (5.38), respectively, can be substituted into (5.30) to obtain the approximate de-embedding error of the *Insertion Loss* technique in decibel

$$\varepsilon\,(\text{dB}) \approx \tilde{G}_{\text{DUT}} - G_{\text{DUT}} = 20\log\left(\frac{|s_{21,\text{meas}}|}{|s_{21,\text{b2b}}| \cdot \left|s_{21,\text{dd}}^{\text{DUT}}\right|}\right), \tag{5.42}$$

where $G_{\text{DUT}} = \left|s_{21,\text{dd}}^{\text{DUT}}\right|^2$ is the actual gain of the DUT.

As can be seen, the error depends both on the properties of the DUT and of the balun. In particular, the error also depends on the magnitude of the transmission S-parameter of the amplifier $s_{21,\text{dd}}^{\text{DUT}}$ that has to be de-embedded. Additionally, after developing (5.42) it can be observed that an error of 0 dB in magnitude is possible only under very restricting conditions on the balun S-parameters

$$s_{\text{dc}} \approx 0,\ s_{1c} \approx 0,\ s_{\text{dd}} \approx 0. \tag{5.43}$$

In practice, this usually requires that these parameters remain below $-30\,\text{dB}$ over the frequency range of interest [14]. Thus, the de-embedded frequency characteristics are usually deformed and cannot be reconstructed using the *Insertion Loss* method.

Another observation can be made without any restricting requirements on the balun parameters, but on the DUT. Assuming that the characterized differential amplifier has a very good port matching

$$s_{11,\text{dd}}^{\text{DUT}} \approx 0,\ s_{22,\text{dd}}^{\text{DUT}} \approx 0, \tag{5.44}$$

a very good reverse isolation

5.2 Differential Measurements using Baluns

$$s_{12,dd}^{DUT} \approx 0, \tag{5.45}$$

and a very good common-mode rejection

$$s_{21,cc}^{DUT} \approx 0, \tag{5.46}$$

the measured transmission S-parameter of the setup in Fig. 5.17 simplifies to

$$s_{21,meas} \approx s_{1d}^2 \cdot s_{21,dd}^{DUT}. \tag{5.47}$$

Thus, only in the case of an "ideal" de-embedded amplifier, the error in (5.42) does not depend on the magnitude of the transmission S-parameter $s_{21,dd}^{DUT}$, but only on the S-parameters of the balun

$$\varepsilon(dB) = 20\log \left| \frac{s_{1d}^2 \cdot [(s_{dd}+1)(s_{cc}+1) - s_{dc}^2] \cdot [(s_{cc}-1)(s_{dd}-1) - s_{dc}^2]}{(1 - s_{dc}^2 - s_{cc}^2) s_{1d}^2 + (1 - s_{dd}^2 - s_{dc}^2) s_{1c}^2 + 2s_{1c}s_{dc}s_{1d}(s_{cc} + s_{dd})} \right|. \tag{5.48}$$

5.2.2 Measurement Verification

The above theoretical considerations have been verified in measurement and simulation. Firstly, two on-board test baluns based on a hybrid ring coupler, shown in Fig. 5.19, have been designed for the center frequency of 24 GHz. Balun *A* has been designed to be symmetrical and balun *B* to be asymmetrical.

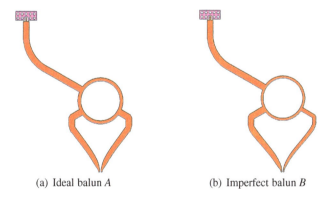

(a) Ideal balun *A*   (b) Imperfect balun *B*

Fig. 5.19 Design of two 24 GHz baluns.

Thus, balun *A* offers good port isolation and matching, whilst balun *B*, has higher differential to common-mode conversion and degraded port matching. One branch of the balun is wider than the other one. Furthermore, the width of the trace on the

70   5 Measurement Techniques

hybrid ring coupler circumference was designed to have an impedance of 82.7 Ω, instead of 70.7 Ω necessary for a 50 Ω system.

A direct characterization of such three-port devices in measurement is not trivial, thus the structures have been accurately simulated using HFSS. In order to obtain insight on the properties of the baluns and whether they fulfill the requirements in (5.43), the simulated modal S-parameters presented in Fig. 5.20 are considered.

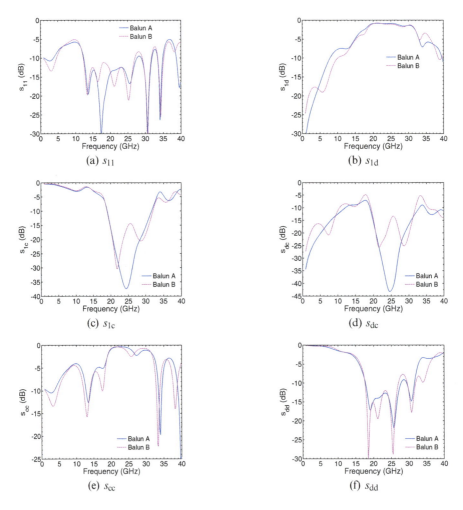

Fig. 5.20 Simulated modal S-parameters of the 24 GHz baluns.

The mode conversion parameters s_{1c} and s_{dc} of the balun A are well below -30 dB in the frequency range $22.5 - 26.5$ GHz. As can be seen, the balun B has much higher mode conversion than balun A, and can be used as an example of a balun that does not fulfill the requirements in (5.43). The differential matching of

5.2 Differential Measurements using Baluns

better than $-20\,\text{dB}$ in the vicinity of the center frequency, achieved by balun A is a good practical value. However, poorer matching further away from the center frequency may be responsible for a higher de-embedding error at these frequencies.

Fig. 5.21 presents a test board, having a back-to-back interconnection of the balun A with its mirrored version A'. This test structure is used for the *Insertion Loss* de-embedding of the balun and corresponds to the setup in Fig. 5.18.

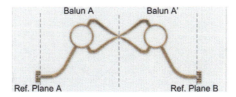

Fig. 5.21 On-board test structure for balun de-embedding.

The structure has been characterized on-board using Cascade Microtech ACP probes up to 40 GHz with 200 μm pitch in GSG configuration. The reference planes have been set using the four-port SOLT calibration.

The back-to-back connection of the balun A has also been simulated in HFSS and compared with the measured S-parameters of the test structure in Fig. 5.21. Furthermore, the simulated modal S-parameters of balun A in Fig. 5.19(a) have been used in (5.36) and (5.37) to obtain the back-to-back S-parameters analytically. The comparison of the results is presented in Fig. 5.22.

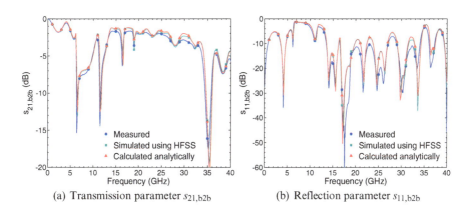

(a) Transmission parameter $s_{21,\text{b2b}}$ (b) Reflection parameter $s_{11,\text{b2b}}$

Fig. 5.22 Comparison of measured, simulated and calculated baluns setup.

As can be observed, due to very careful modeling of the PCB materials and very dense meshing in HFSS, a very good match between measured and simulated results is achieved. Thus, the equations (5.36) and (5.37) have been verified.

The comparison of the good balun *A* versus the poor balun *B* allows to evaluate the error due to balun's imperfect properties. The simulated S-parameters of both baluns have been used in (5.48) to estimate the error due to *Insertion Loss* de-embedding of an "ideal" amplifier, as presented in Fig. 5.23.

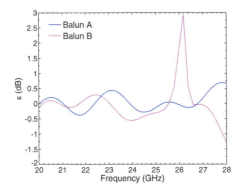

Fig. 5.23 Comparison of the de-embedding error due to balun parameters.

The error of the balun *B* reaches a maximum of 2.9 dB at frequency of 26.2 GHz. This is due to the peaks in common-mode transmission s_{1c} and mode conversion s_{dc} S-parameters, reaching a maximum of -15 dB at this frequency in Fig. 5.20(c) and Fig. 5.20(d), respectively. The error can also become negative, which means the measured insertion loss of a balun can be estimated lower than it actually is.

Secondly, a 24 GHz differential LNA, presented in [24] and described in detail in section 6.1.2 has been used as a DUT. The LNA chip has been mounted on a board with baluns, as presented in Fig. 5.24.

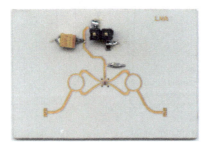

Fig. 5.24 Test board for two-port characterization of the LNA.

The baluns are realized using a hybrid ring coupler centered at 24 GHz, similar to balun *A* in Fig. 5.19(a). Furthermore, a chip realizing a thru standard has been bonded on an additional similar board. This allows the connection of the error

5.2 Differential Measurements using Baluns

boxes consisting of the balun, bondwires and a short on-chip microstrip line with its mirrored version back-to-back. Obviously, a further minor inaccuracy may be introduced by non-identical error boxes caused by poor repeatability of the bondwires. However, this inevitable inaccuracy of the described error network is minimal compared to a typical laboratory coaxial measurement setup, using for example off-the-shelf hybrid couplers.

The two-port measurements have been performed on-board using GSG configuration. The direct four-port measurement has been performed on-wafer using Cascade Microtech Infinity probes up to 40 GHz with 100 µm pitch in GSSG configuration. Again, four-port SOLT calibration has been performed. The nodal S-parameters are converted into modal and used for comparison.

The measured two-port S-parameters of the DUT $s_{21,\text{meas}}$ and of the back-to-back connection of the baluns $s_{21,\text{b2b}}$ have been used in (5.30) and compared with the directly measured results using a four-port VNA. The comparison of the magnitude of the differential transmission S-parameter is presented in Fig. 5.25.

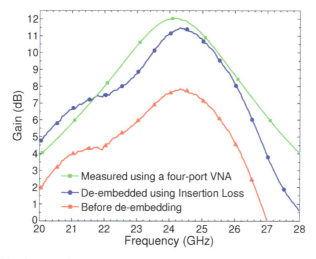

Fig. 5.25 LNA gain comparison.

As can be observed, there is a gain deviation and a minor frequency shift in the de-embedded gain characteristics. Furthermore, there is a shape deformation of the frequency characteristics of the LNA. An additional erroneous inflection point is observed at 22 GHz for the curve de-embedded using the *Insertion Loss* method. A maximum deviation of several decibels is observed in Fig. 5.25 between the directly measured and de-embedded results.

References

1. L. Martens, *High-frequency characterization of electronic packaging*, chapter 3.5.3, Springer, 1998.
2. M. C. A. M. Koolen, J. A. M. Geelen, and M. P. J. G. Versleijen, "An Improved De-embedding Technique for On-Wafer High-Frequency Characterization", *in Proc. Bipolar / BiCMOS Circuits and Technology Meeting (BCTM)*, pp. 188--191, Minneapolis, USA, September 1991.
3. L. F. Tiemeijer, R. J. Havens, A. B. M. Jansman, and Y. Bouttement, "Comparison of the "Pad-Open-Short" and "Open-Short-Load" deembedding techniques for accurate on-wafer rf characterization of high-quality passives", *IEEE Transactions on Microwave Theory and Techniques*, vol. 51, pp. 723--729, February 2005.
4. Y. Tretiakov, K. Vaed, W. Woods, S. Venkatadri, and T. Zwick, "A New On-Wafer De-Embedding Technique for On-Chip RF Transmission Line Interconnect Characterization", *in Proc. IEEE 63rd ARFTG Conference*, pp. 69--72, Fort Worth, USA, June 2004.
5. A. Issaoun, Y. Z. Xiong, J. Shi, J. Brinkhoff, and F. Lin, "On the Deembedding Issue of CMOS Multigigahertz Measurements", *IEEE Transactions on Microwave Theory and Techniques*, vol. 55, pp. 1813--1823, September 2007.
6. J. Song, F. Ling, G. Flynn, W. Blood, and E. Demircan, "A De-embedding Technique for Interconnects", *in Proc. IEEE Elec. Perf. of Electronic Packaging*, pp. 129--132, Cambridge, USA, October 2001.
7. M. B. Steer, S. B. Goldberg, G. Rinne, P. D. Franzon, I. Turlik, and J. S. Kasten, "Introducing the through-line deembedding procedure", *in IEEE MTT-S International Microwave Symposium (IMS) Digest*, pp. 1455--1458, Albuquerque, USA, June 1992.
8. G. F. Glenn and C. A. Hoer, "Thru-reflect-line: An improved technique for calibrating the dual six-port automatic network analyzer", *IEEE Transactions on Microwave Theory and Techniques*, vol. 27, pp. 987--993, December 1979.
9. V. Issakov, M. Wojnowski, A. Thiede, and L. Maurer, "Extension of Thru De-embedding Technique for Asymmetrical and Differential Devices", *IET Circuits, Devices & Systems*, vol. 3, pp. 91--98, April 2009.
10. D. E. Bockelman and W. R. Eisenstadt, "Combined Differential and Common-Mode Scattering Parameters: Theory and Simulation", *IEEE Transactions on Microwave Theory and Techniques*, vol. 43, pp. 1530--1539, July 1995.
11. T. Zwick and U. Pfeiffer, "Pure-mode network analyzer concept for on-wafer measurements of differential circuits at millimeter-wave frequencies", *IEEE Transactions on Microwave Theory and Techniques*, vol. 53, pp. 934--937, March 2005.
12. C. Seguinot, P. Kennis, J.-F. Legier, F. Huret, E. Paleczny, and L. Hayden, "Multimode TRL - A New Concept in Microwave Measurements: Theory and Experimental Verification", *IEEE Transactions on Microwave Theory and Techniques*, vol. 46, pp. 536--542, May 1998.
13. D.-H. Han, T. Q. Ruttan, and L. A. Polka, "Differential de-embedding methodology for on-board CPU socket measurements", *in Proc. IEEE 61st ARFTG Conference*, pp. 37--43, Philadelphia, USA, June 2003.
14. V. Issakov, M. Wojnowski, A. Thiede, and R. Weigel, "Considerations on the De-embedding of Differential Devices Using Two-Port Techniques", *in European Microwave Conference (EuMC)*, pp. 695--698, Rome, Italy, October 2009.
15. A. G. Chiariello, A. Maffucci, G. Miano, F. Villone, and W. Zamboni, "A Transmission-Line Model for Full-Wave Analysis of Mixed-Mode Propagation", *IEEE Transactions on Advanced Packaging*, vol. 31, pp. 275--284, February 2008.
16. O. Zinke and H. Brunswig, *Hochfrequenztechnik 1*, chapter 4.11, Springer Verlag, 6th edition, 2000.
17. R. B. Marks, "A multiline method of network analyzer calibration", *IEEE Transactions on Microwave Theory and Techniques*, vol. 39, pp. 1205--1215, July 1991.
18. D. Pozar, *Microwave Engineering*, Wiley, 2nd edition, 1998.

References 75

19. J. C. Tippet and R. A. Speciale, "A Rigorous Technique or Measuring the Scattering Matrix of a Multiport Device with a 2-port Network Analyzer", *IEEE Transactions on Microwave Theory and Techniques*, vol. 30, pp. 661--666, May 1982.
20. S. Belkin, "Differential Circuit Characterization with Two-Port S-Parameters", *IEEE Microwave Magazine*, vol. 7, pp. 86--99, December 2006.
21. M. Spirito, M. P. van der Heijden, M. de Kok, and L. C. N. de Vrede, "A calibration procedure for on-wafer differential load-pull measurements", *in Proc. IEEE 61st ARFTG Conference*, pp. 1--4, Philadelphia, USA, June 2003.
22. V. A. Monaco and P. Tiberio, "Computer-Aided Analysis of Microwave Circuits", *IEEE Transactions on Microwave Theory and Techniques*, vol. 22, pp. 249--263, March 1974.
23. K. C. Gupta, R. Garg, and R. Chadha, *Computer-Aided Design of Microwave Circuits*, chapter 11.2.2, Artech House, 1981.
24. V. Issakov, H. Knapp, M. Wojnowski, A. Thiede, W. Simbürger, G. Haider, and L. Maurer, "ESD-protected 24 GHz LNA for Radar Applications in SiGe:C Technology", *in Topical Meeting on Silicon Monolithic Integrated Circuits in RF Systems (SiRF)*, pp. 1--4, San Diego, USA, January 2009.

Chapter 6
Radar Receiver Circuits

Even though MOSFET has been invented much earlier [1] than bipolar transistor [2], the commercial mass-volume foundry implementation of bipolar technology has been introduced more than a decade prior to CMOS. Therefore, some of the classical analog circuit topologies e.g. Gilbert cell mixer or current-mode logic (CML), originally developed for bipolar transistors, were directly adopted in CMOS. However, CMOS technology has differing characteristics that have to be considered during circuit design. Some properties can be utilized to gain advantages. For example, true CMOS logic circuits consume much less current than their CML counterparts. On the other hand, MOS transistors unlike bipolar suffer from high 1/f noise and may hinder straightforward implementation of the Gilbert cell mixer topology for the direct down-conversion architecture. Several circuit techniques have been developed in CMOS in order to overcome this problem [3], [4]. An additional common approach, adopted from III/V technologies, is to use passive resistive mixers [5]. This technique is well suited for MOSFETs, since these may act as passive voltage-controlled switches, but it is not applicable for bipolar transistors since these act as current-controlled switches.

This chapter presents the design, realization, and characterization of 24 GHz narrow-band receiver circuits in Infineon's C11N Si CMOS and B7HF200 SiGe:C HBT technologies, described in detail in chapter 3. For the sake of simplicity, these technologies are further briefly referred to as *CMOS* and *SiGe*. Receiver building blocks LNA and mixer have been designed in both technologies and compared in sections 6.1 and 6.2, respectively. Different circuit topologies are considered in each case. Additionally, a detailed systematic analysis of system considerations using the example of CMOS receivers consisting of LNA with active mixer versus LNA with passive mixer is presented in section 6.3. The building blocks are integrated into IQ receivers in CMOS and SiGe and described in section 6.4. Section 6.5 presents compact integrated 90° and 180° power splitters/combiners and a hybrid ring coupler. Finally, section 6.6 describes circuit-level ESD protection concepts for RF circuits analyzed in this work.

V. Issakov, *Microwave Circuits for 24 GHz Automotive Radar in Silicon-based Technologies*, DOI 10.1007/978-3-642-13598-9_6, © Springer-Verlag Berlin Heidelberg 2010

6.1 Low-Noise Amplifiers

There is a large amount of circuit topologies and techniques for the design of low-noise amplifiers, described in numerous publications [6], [7]. The most widely used topologies are common-source (CS) and common-gate (CG) for CMOS or common-emitter (CE) and common-base (CB) for bipolar. An additional development is merging these into a cascode structure. The amplification is then performed using the transistor in CS or CE configuration, whilst the transistor with CG or CB is used to offer a higher load impedance and to isolate the output from the input.

Section 6.1.1 presents a 24 GHz current-reusing CMOS LNA proposed in this work and described in [8]. Section 6.1.2 presents a cascode 24 GHz SiGe LNA designed in this work and described in [9]. Section 6.1.3 presents the measurement results achieved by both circuits. Finally, section 6.1.4 summarizes, compares and analyzes the performance of both circuits.

6.1.1 LNA in CMOS Technology

The conceptual schematic diagram of the proposed LNA implementation is presented in Fig. 6.1. This is a self-biased pseudo-differential AC coupled two-stage

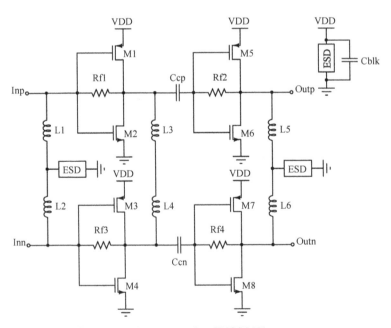

Fig. 6.1 Schematic of the proposed current-reusing CMOS LNA.

6.1 Low-Noise Amplifiers

amplifier. It is based on a combined N-/PMOSFET transconductor reusing DC-current from the supply. The classical inductive source degenerated topology [10] has been disregarded in this case due to low quality factors of inductors, as shown in section 3.3.2.

The combined N-/PMOSFET structure resembles a classical digital inverter circuit. The feedback resistors Rf1, Rf2, Rf3 and Rf4 having large values of 15 kΩ are used for operating point self-biasing by setting the input DC voltage V_{in} to be nearly equal to the output DC voltage V_{out}. If one compares this bias point with the "switching point" on the transfer characteristics of an inverter, one observes that in this case it cannot be at half of the supply voltage, since this would require a large PMOS device, which would considerably affect the high-frequency performance due to the excessive capacitive parasitics.

In order to reduce the capacitive loading by the slower PMOS device, but still to take advantage of its additional transconductance in the signal path, the PMOSFET width has been set to be smaller than the NMOSFET width. The choice of the widths for the input transistors poses a trade-off between the frequency of operation, NF, gain, current consumption and input matching. For high frequency of operation, a small input capacitance is required. Reducing the latter by choosing smaller widths of the input stage transistors results in a higher contribution of the vertical resistance, which is the contact resistance between silicide and gate polysilicon [11], to the overall gate resistance given in Eq. (3.1). Thus, the width is optimized for lowest gate resistance. A higher gate resistance would cause higher white noise contribution and degrade the noise factor, defined as follows

$$F = F_{\text{min}} + \frac{R_n}{G_{\text{src}}} \left| Y_{\text{src}} - Y_{\text{src,opt}} \right|^2 , \qquad (6.1)$$

by causing a higher equivalent noise resistance R_n, given for the CS configuration by [12]

$$R_n = R_s + R_g + R_{n0} , \qquad (6.2)$$

where F_{min} is the minimal noise factor, $Y_{\text{src,opt}}$ is the optimal source impedance for minimum noise figure, Y_{src} and G_{src} are the source admittance and conductance, R_s and R_g are the series parasitic resistances at the MOSFET source and gate terminals and R_{n0} is the noise resistance due to an equivalent input-referred current source, describing the channel thermal noise.

The DC bias point is defined by the choice of NMOS and PMOS transistor sizes. To have maximal f_T and higher g_m all the MOS lengths are to set to 0.13 μm, the minimal allowed length of the technology. Thus, only the optimal transistor widths have to be determined during the bias design.

For simplicity of the following DC bias point analysis, the short channel effects have been neglected. Thus, considering the saturation region of the transistor characteristics, the self-bias point of a single N-/PMOSFET structure is given by

$$V_{in} = V_{out} = \frac{V_{sup} - |V_{thp}| + \sqrt{\frac{\mu_n W_n}{\mu_p W_p}} V_{thn}}{\sqrt{\frac{\mu_n W_n}{\mu_p W_p}} + 1},$$ (6.3)

where V_{in} and V_{out} are the DC voltage levels at the input and output of the structure, V_{thp}, V_{thn}, μ_p, μ_n, W_p, W_n are threshold voltage, mobility and width of N- and PMOSFET, respectively, and V_{sup} is the supply voltage level.

As described in section 3.3.1, there is an optimal current density at which a MOSFET should be biased to achieve the minimal F_{min}. In case of the NLVT in C11N technology, described in section 3.3.1, it has been simulated to be $J_{opt} = 0.1\,mA/\mu m$, which is close to the empirically predicted value reported in [13]. Therefore, the optimal ratio of W_n and W_p for the lowest F_{min} can be easily determined by setting the required DC bias current $I_{opt} = J_{opt} \cdot W_n$ into the well-known equation for the MOSFET drain current in saturation

$$I_d\,[A] = \frac{\mu_n C_{ox} W_n}{2L} (V_{in} - V_{thn})^2 = 100\,[A/m] \cdot W_n\,[m]$$ (6.4)

and substituting (6.3) for the input DC voltage

$$\frac{W_n}{W_p} = \frac{\mu_p}{\mu_n} \left((V_{sup} - |V_{thp}| - V_{thn}) \sqrt{\frac{\mu_n C_{ox}}{200 \cdot L}} - 1 \right)^2.$$ (6.5)

Substituting typical values for the 0.13 µm CMOS technology leads to ratios of about 8-9. This means that in order to set the self-bias for the minimum noise it is required to either have a very small W_p, which would cause high gate resistance, or to set large W_n, which would affect high-frequency performance due to parasitics. Therefore, it is more cumbersome to achieve the minimal F_{min} of the MOSFET using this topology. However, since the actual deviation from the minimal F_{min} for practical ratios of about 1.5-3 is small, the transistor widths are optimized for easier input matching and lowest gate resistance, as mentioned previously.

As described in section 3.1, bulk terminals of all transistors are shorted to sources. The latter are connected in this circuit to GND and VDD for NMOS and PMOS devices, respectively. For the small-signal analysis of the circuit the external transistor parasitics, described in detail in section 3.1.1, are also taken into account. Fig. 6.2 presents the simplified small-signal equivalent circuit of a single inverter in the first stage of the LNA, consisting of a single N-/PMOSFET transconductance pair and inductors between the input and output nodes to the virtual ground, as e.g. transistors M1 and M2 and inductors L1 and L3 in Fig. 6.1. The physical models of the on-chip spiral inductors at the input and output of the stage are simplified in the first approximation by equivalent series resistances R_{sin}, R_{sout} and inductances L_{in}, L_{out} in Fig. 6.2. The inductors are connected to the virtual AC ground at the symmetry axis. Unfortunately, the gate resistances of transistors cannot be neglected. The indices n or p denote N- or PMOSFET, respectively. The transconductances $g_{mn,p}$ are multiplied by the voltages $v_{n,p}$, which are related to the small-signal input voltage

6.1 Low-Noise Amplifiers

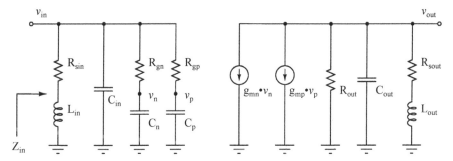

Fig. 6.2 Simplified small-signal equivalent circuit of a single inverter stage.

v_{in} by the voltage divider between the gate resistance and the input capacitances as follows

$$v_{n,p} = \frac{1}{1 + j\omega R_{gn,p} C_{n,p}} v_{in}. \qquad (6.6)$$

The overall voltage gain between the input and output nodes in Fig. 6.2 is denoted by A_t. According to Miller's theorem [14], any feedback impedance Z_f connected between the input and output nodes of an amplifier, can be replaced by equivalent input and output impedances $Z_f(1-A_t)$ and $Z_f(1-1/A_t)$, respectively. The gain A_t of the single inverter stage is assumed to be sufficiently high, such that the Miller impedance at the output is approximately equal to the feedback impedance Z_f. In the analyzed circuit, the feedback path between input and output nodes of the structure comprises intrinsic $C_{gdn,p}$ and layout-dependent external $C_{gdmn,p}$ parasitic capacitances of the N-/PMOS transistors, and a feedback resistor R_f, corresponding to Rf1 or Rf2 in Fig. 6.1. These elements appear at the output by approximation, whilst at the input they are seen multiplied by the factor $(1-A_t)$. The feedback resistor has been neglected in Fig. 6.2, since R_f has a large value, which is further increased due to the Miller effect.

The components in the simplified equivalent circuit in Fig. 6.2 are composed of several contributions. The capacitors C_{in} and C_{out} describe the input and output capacitances of the inverter, the capacitors $C_{n,p}$ represent the input capacitances of each transistor and the resistor R_{out} describes the total output resistance. These equivalent components can be explicitly written as

$$C_{in} = C_{gsmn} + C_{gsmp} + (C_{gdmn} + C_{gdmp})(1-A_t), \qquad (6.7)$$

$$C_{n,p} = C_{gsn,p} + C_{gdn,p}(1-A_t), \qquad (6.8)$$

$$C_{out} \approx C_{dbp} + C_{gdp} + C_{dbn} + C_{gdn} + C_{dsmn} + C_{dsmp} + C_{gdmn} + C_{gdmp} + C_l, \qquad (6.9)$$

$$R_{out} \approx r_{dsn} \| r_{dsp} \| R_f \| R_l, \qquad (6.10)$$

where $C_{gsmn,p}, C_{gdmn,p}$ and $C_{dsmn,p}$ describe external parasitics due to the finger layout metallization, as described in Fig. 3.2. The terms $C_{gsn,p}, C_{gdn,p}$ and $C_{dbn,p}$ describe the equivalent gate-source, gate-drain and drain-bulk internal transistor parasitic capacitances. The terms $r_{\text{dsn,p}}$ model the finite transistor drain-source resistances. Additionally, the external load provided by the following stage is modelled by resistor R_l and capacitor C_l. The overall frequency-dependent voltage gain A_t of the single inverter stage in Fig. 6.2 is given by

$$A_t = -\left(\frac{g_{\text{mn}}}{1 + j\omega R_{\text{gn}}C_n} + \frac{g_{\text{mp}}}{1 + j\omega R_{\text{gp}}C_p}\right)\left(R_{\text{out}}\|\frac{R_{\text{sout}} + j\omega L_{\text{out}}}{1 - \omega^2 L_{\text{out}}C_{\text{out}} + j\omega R_{\text{sout}}C_{\text{out}}}\right).$$
$$(6.11)$$

The impact of the high value feedback resistance R_f on the overall gain is negligible, as can be estimated from (6.10) and (6.11). It can be further observed that if the gate resistances were negligible, the transconductance would have become $g_{\text{mn}} + g_{\text{mp}}$, corresponding to the DC value. This means that the gate resistances R_{gn}, R_{gp} along with the input capacitances C_n, C_p compose a low-pass filter that deteriorates the high-frequency gain. Thus, the gate resistance has to be minimized to reduce the filter constant and accordingly to increase the corner frequency.

The inductance at the output of the stage resonates off the output capacitance of this stage and the input capacitance of the following stage at the center frequency. Therefore, the parasitic capacitances of the stages as well as of the load are compensated by the on-chip inductors L1-L6 at the center frequency. Assuming that in the vicinity of the center frequency the real part of the last terms' denominator is null, thus the gain in (6.11) can be rewritten and further simplified by assuming that the quality factor of the inductor at the output Q_{out} is much higher than one

$$A_t \approx -\left(\frac{g_{\text{mn}}}{1 + j\omega R_{\text{gn}}C_n} + \frac{g_{\text{mp}}}{1 + j\omega R_{\text{gp}}C_p}\right)\left(R_{\text{out}}\|\frac{Q_{\text{out}}}{\omega C_{\text{out}}}\right).$$
$$(6.12)$$

Thus the gain of the first stage depends on the quality factor of the inductors at the output of the first stage, L3, L4 in Fig. 6.1. The gain of the second stage also depends on the inductors at its output L5, L6. However, it might be limited by the low load resistance.

Next, the input impedance of the inverter stage in Fig. 6.2 can be written as

$$Z_{\text{in}} = \left(R_{\text{gn}} + \frac{1}{j\omega C_n}\right)\|\left(R_{\text{gp}} + \frac{1}{j\omega C_p}\right)\|\left(\frac{R_{\text{sin}} + j\omega L_{\text{in}}}{1 - \omega^2 L_{\text{in}}C_{\text{in}} + j\omega R_{\text{sin}}C_{\text{in}}}\right). \quad (6.13)$$

As can be observed, the input impedance strongly depends on the gate resistance of the transistors and the input capacitance. For sufficiently small gate resistance the first two terms can be simplified into $1/\left(j\omega\left(C_n + C_p\right)\right)$, representing an equivalent input capacitance. The last term in (6.13) can be again simplified under assumption of quality factor much higher than one and Eq. (6.13) simplifies in the vicinity of the center frequency to the following form

$$Z_{\text{in}} \approx \left(R_{\text{gn}} + \frac{1}{j\omega C_n}\right)\|\left(R_{\text{gp}} + \frac{1}{j\omega C_p}\right)\|\left(\frac{Q_{\text{in}}}{\omega C_{\text{in}}}\right), \quad (6.14)$$

6.1 Low-Noise Amplifiers

where C_{in} is given in (6.7) and Q_{in} is the quality factor of the input inductor, L1 or L2 in Fig. 6.1. Thus, the input impedance depends on the quality factor of the input inductors. The LC tank is used to set the real part of the input and the output impedances. Additionally, for lower capacitance better input matching is achieved.

To sum up, there is a trade-off between gain, noise figure and matching, which has to be optimized through an appropriate choice of the transistor widths. Larger MOSFET width contributes to higher transconductance since in saturation holds [14]

$$g_m = \mu_n C_{ox} \frac{W}{L} \left(V_{gs} - V_t \right). \qquad (6.15)$$

From Eq. (6.11), higher transconductance results in higher gain. However, larger width increases parasitic capacitance and thus deteriorates the high-frequency performance. On the other hand, smaller width increases vertical gate resistance in (3.1) and thus the overall noise figure in (6.1), but makes input impedance matching easier. Thus, the optimal values of the widths have been found to be 25 µm for PMOSFET and 40 µm for NMOSFET, respectively.

It is assumed that the LNA is excited differentially, whilst the 180° phase shift is generated externally. The symmetry axis introduces a constant potential for differential mode variations, thus it can be considered as a virtual AC ground. This means, inherently highly capacitive devices such as ESD protection structures, added at the symmetry line of the chip, would be transparent to the high-frequency signals and would not impact the circuit performance, as explained in detail in section 6.6.2. In this circuit the ESD devices have been attached between two inductors, as shown in Fig. 6.1. Furthermore, if the layout concept allows for sufficient area, the same virtual ground concept can be used to attach further devices such as bypass capacitors. A capacitor attached in the symmetry axis between two inductors provides a series LC resonant path to ground for common-mode signals. The resonance frequency can be set by the choice of the capacitor value to a lower frequency of few gigahertz, at which the amplifier has a high common-mode gain. At the resonance frequency, the LC branch shorts the common-mode signals to ground and introduces a notch in the transmission characteristics, thereby improving the common-mode stability of the circuit.

In the presented design the capacitors between the stages C_{cn} and C_{cp} are used to separate the self-biased DC operation points of the amplifier stages. Additionally, the bypass capacitor Cblk has been added between VDD and GND to provide a low impedance path between the supplies at high frequencies.

6.1.2 LNA in SiGe:C Technology

The conceptual schematic diagram of the LNA is presented in Fig. 6.3. It is a differential cascode LNA with inductive emitter degeneration and parallel tanks as loads. The cascode topology is very popular in bipolar circuits due to the high HBT transconductance. This topology combines the high gain, good linearity and ease of matching offered by a common-emitter stage with the improved isolation provided by a common-base stage.

Sufficient gain can be achieved due to high transconductance offered by the transistors and high quality factors of inductors used in the LC tank. Due to power consumption and linearity considerations, the LNA has been realized using a single amplifying stage. As has been shown in section 3.3.2, the B7HF200 technology offers considerably higher quality factors compared to C11N, thus making this process more suitable for this narrow-band circuit implementation.

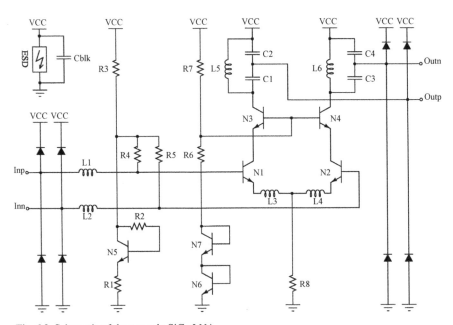

Fig. 6.3 Schematic of the cascode SiGe LNA.

In contrast to the circuit topologies presented in [15] and [16], a common-collector stage for output matching is avoided here. Instead, reactive impedance transformation is used, since it offers improved linearity and stability. The impedance transformation is realized by a tapped tank capacitance, providing a reactive voltage divider between C1, C2 and C3, C4, respectively. This allows down-transformation of the real part of the tank impedance to have a better match to 50 Ω at the output. The simplified calculation is straightforward [17]

$$\frac{R_{\text{out}}}{R_{\text{in}}} \approx \left(\frac{C1}{C1+C2}\right)^2. \tag{6.16}$$

The LC tanks including also L5 and L6 are resonant at the center frequency of 24 GHz. They increase the gain by introducing high-resistance loads and improve the output matching by tuning out the pad capacitances, which are included in the

6.1 Low-Noise Amplifiers

tanks. Moderately high quality factors of inductors in B7HF200 allow achieving sufficient gain, without further concerns on the stability of the circuit.

The transistors N3 and N4 are used to reduce the Miller effect at the base-collector capacitances C_{bc} of N1 and N2. The implementation of a cascode also provides higher gain due to larger output impedance, but has the disadvantage of reducing the headroom and thus affecting the linearity. Therefore, the LNA gain has been adjusted to be not too high to allow for good linearity, necessary for a receiver chain that requires an overall IP1dB > -15 dBm, but still sufficient to provide an overall conversion gain of at least 9 dB over the required range of temperatures.

In this design the input stage consisting of transistors N1, N2 has been optimized for low noise figure following the procedure presented in [7]. Firstly, the transistors have been biased at the optimum current density that minimizes transistor noise figure F_{min}. As shown in Fig. 3.10(b), for a high-speed npn transistor in B7HF200 the minimal noise figure at 24 GHz is achieved at a current density around 0.6 mA/μm^2, which is significantly below the optimum current density for peak f_T and g_m. Thus, biasing for low noise figure compromises gain.

Secondly, the transistors N1, N2 have been dimensioned such that the real part of the optimum noise admittance $Y_{src,opt}$ in Eq. (6.1) is equal to the real part of the source admittance at 24 GHz. Larger devices are required at this frequency to achieve a good match, resulting in increased current consumption. However, larger devices are also advantageous, since they offer improved linearity due to higher base-collector capacitance C_{bc} and lower base resistance r_b [18]. Furthermore, as shown in [19], the base resistance of a SiGe HBT is directly proportional to the noise resistance R_n and given by

$$R_n = r_b + \frac{1}{2g_m}, \tag{6.17}$$

where g_m is the transistor transconductance. Usually $r_b > 1/2g_m$, which leads to $R_n \approx r_b$. Therefore, r_b should be minimized in order to keep the noise factor in Eq. (6.1) close to F_{min} in case that the optimal noise impedance is not well matched. Additionally, the minimum noise factor F_{min} is also related to the base resistance r_b and is given in general form as follows [19]

$$F_{min} = 1 + \frac{1}{\beta} + \sqrt{\frac{2g_m r_b + 1}{\beta} + \frac{r_b \left(\omega (C_{be} + C_{bc}) \right)^2}{g_m}}, \tag{6.18}$$

where β is the DC current gain of the transistor. In order to achieve an optimal F_{min}, a small base resistance r_b, high current gain β and small transit times equivalent to small parasitic capacitances C_{be}, C_{bc} are required. The optimal emitter length of N1, N2 was found to be 16.4 μm. These transistors have been laid-out in the double-base contact configuration CBEBEBC, similar to the description in section 3.2.1, in order to reduce the base resistance r_b.

Next, the input impedance has been tuned by the classical approach of adding series inductances L1, L2 at the bases and inductive degeneration by L3, L4 at the emitters. The values of these base and emitter inductors are chosen using the well-known expressions for simultaneous noise and impedance match [20]

$$L3 = L4 = \frac{Z_0}{2\pi f_T}, \tag{6.19}$$

$$L1 = L2 = \frac{1}{\omega^2 C_{\text{in}}} - L3, \tag{6.20}$$

where Z_0 is the source resistance and C_{in} is the input capacitance of the transistor.

The bias network for the input transistors is based on a current mirror using transistor N5 in the reference branch. The total current in the transistors N1 and N2 is four times the reference branch current. Resistors R2 and R4, R5 maintain the mirror ratio of 4:1. Resistors R4 and R5 have large values in order to separate the bias network from the RF signal. Resistor R2 compensates for the asymmetry introduced due to the voltage drop on R4 and R5 in the current mirror. The current mirror compensates for finite β using the feedback around transistor N5. In order to improve the stability over process variations and tolerances, matching of the transistors N1, N2 to N5 and resistors R4, R5 to R2 as well as R1 to R2 has been carefully taken into consideration.

The bias network for the cascode transistors N3 and N4 is considerably simpler. It consists of two diode-connected transistors N6 and N7 and the resistive divider between R6 and R7, which sets the base potential of N3 and N4. The choice of the values for R6 and R7 poses a tradeoff between low power dissipation and low common-mode impedance at the base of the cascode transistors.

Additionally, ESD protection diodes have been attached at the RF pins. The devices are optimized for lowest capacitance and smallest area, in order to minimize deterioration of the high-frequency performance, but still to maintain highest possible hardness. Diode ESD structures have the advantage of faster switching than snap-back devices, thus providing better protection for ESD pulse events with faster rise times. However, in order to achieve high protection levels, large area devices are required. Furthermore, power clamping devices have been attached between VCC and GND.

6.1.3 Measurements of CMOS and SiGe LNAs

The circuits, described in sections 6.1.1 and 6.1.2, have been processed in Infineon's C11N and B7HF200 technologies, respectively. The annotated chip micrographs of the CMOS and SiGe LNA circuits are presented in Fig. 6.4. The size including the pads is 0.23 mm^2 for CMOS and 0.32 mm^2 for the SiGe chip, respectively.

The CMOS chip was thinned to 185 µm and the SiGe chip to 350 µm in order to keep the bondwires as short as possible. Both chips have been mounted on a test board, shown in Fig. 6.5. All the measurements have been performed on-board, apart from on-wafer S-parameter measurements of the SiGe LNA.

6.1 Low-Noise Amplifiers

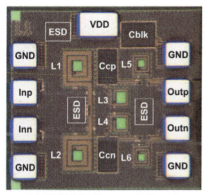
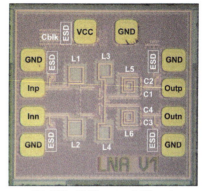

(a) CMOS (0.47 mm × 0.49 mm) (b) SiGe (0.59 mm × 0.55 mm)

Fig. 6.4 Chip micrographs of the LNA circuits.

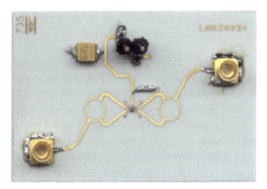

Fig. 6.5 LNA characterization board.

The single-ended to differential conversion has been realized by hybrid ring couplers on the board. The de-embedding has been performed using the *Insertion Loss* technique by connecting the baluns back-to-back. The measurement setup losses have been carefully accounted for using power sensors.

The CMOS LNA consumes 12 mA from 1.5 V, whilst the SiGe LNA consumes 12.6 mA from a single 3.3 V supply.

The measured gain and NF of the CMOS LNA after de-embedding are presented in Fig. 6.6. The deformation of the measured frequency characteristics compared to simulation is due to inaccuracy of balun de-embedding, as analyzed in detail in section 5.2. In the narrow frequency band around the center frequency, where the baluns exhibit ideal 180° phase shift, the de-embedded and the simulated results agree very well. The good match is due to the accurate modeling of the parasitics in the design stage, as described in detail in section 3.1.1. The CMOS LNA achieves a gain of 14 dB and an NF of 5 dB at the center frequency of 24 GHz.

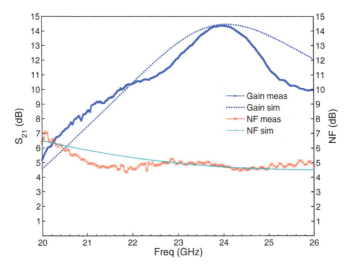

Fig. 6.6 Measured and simulated gain and noise figure of CMOS LNA.

The on-wafer measurement of the SiGe LNA has been performed using Agilent's network analyzer E8364A with a multiport test set Z5623A up to 50 GHz and Cascade Microtech Infinity probes up to 40 GHz with 100 μm pitch in GSSG configuration. Accurate 4-port SOLT (Short-Open-Load-Thru) calibration has been performed to set the reference planes at the input and output of the LNA. The measured four-port nodal S-parameters have been converted into modal representation and the differential gain and output matching are presented in Fig. 6.7. Good match between the measured and simulated results has been achieved due to careful modeling of the parasitics and accurate on-wafer measurement.

The on-board noise figure measurement and simulation are presented in Fig. 6.8. As can be seen, a minimum NF of 3.1 dB is obtained at the center frequency of 24 GHz and the noise figure remains below 3.3 dB over the ISM band, in the frequency range 24 – 24.25 GHz.

The input-referred P1dB (IP1dB) and input-referred third-order intercept point (IIP3) linearity measurements have been performed for both chips at 24 GHz. The IIP3 has been measured by applying two tones at 24 GHz and 24.010 GHz to the LNA input and observing the third order non-linear products at 24.009 GHz and 24.011 GHz at the LNA output. Fig. 6.9 and 6.10 present the IP1dB and IIP3 measurements, respectively, of the CMOS and SiGe LNA circuits. The CMOS LNA achieves an IP1dB and an IIP3 of −14.1 dBm and −1.7 dBm, respectively, whilst the SiGe LNA achieves −8.7 dBm and −1.8 dBm.

The ESD robustness at the RF pins has been tested for both chips using the approach [21]. The I/V and spot leakage curves of the ESD TLP stress between RF input and GND with 100 ns wide pulses having rise time of 10 ns is presented in Fig. 6.11(a) for the CMOS LNA and in Fig. 6.11(b) for the SiGe LNA, respectively. The CMOS chip achieves the lowest TLP failure current of 2 A, whilst the SiGe

6.1 Low-Noise Amplifiers

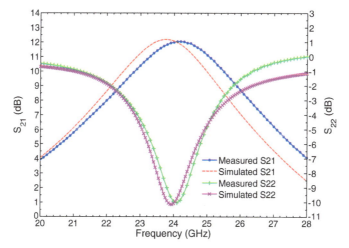

Fig. 6.7 Measured and simulated gain and output matching of SiGe LNA.

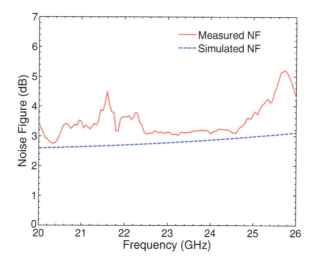

Fig. 6.8 Measured and simulated noise figure of SiGe LNA.

chip has a failure current of 1.7 A on the RF pins. The higher ESD protection level of the CMOS circuit is due to the implementation of the virtual ground principle, described in detail in section 6.6.2. In the SiGe LNA circuit topology, shown in Fig. 6.3, this approach is not straightforward, since the circuit lacks shunt inductors between the differential nodes. However, it might be still realized by attaching a shunt inductor between bases of N1 and N2 and adding the ESD device at the center tap. This would not impact the bias point, since both nodes are at the same DC potential.

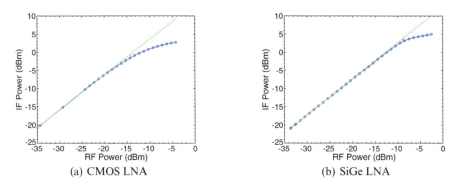

Fig. 6.9 IP1dB measurement of CMOS and SiGe LNA.

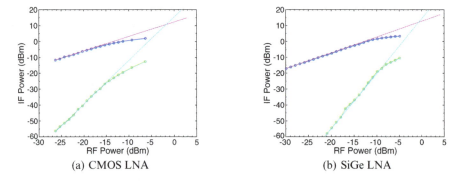

Fig. 6.10 IIP3 measurement of CMOS and SiGe LNA.

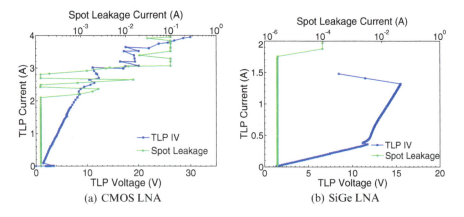

Fig. 6.11 TLP stress measurement of the LNA input to GND.

The robustness of both LNAs has been verified over a wide range of temperatures from $-25\,°C$ to $125\,°C$. Fig. 6.12 presents gain and noise figure as a function

6.1 Low-Noise Amplifiers

of temperature. As expected, at lower temperatures the performance is generally im-

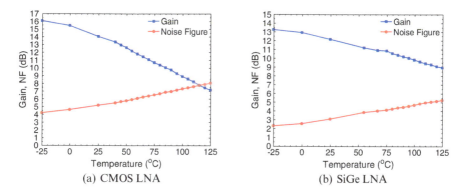

Fig. 6.12 Measured LNA gain and noise figure as function of temperature.

proved. From $-25\,°C$ to $125\,°C$ the gain drops by 9 dB for CMOS and by 4.2 dB for SiGe. The noise figure increases by 3.8 dB and 2.8 dB, respectively. Following the considerations in [9], the major impact on the high-temperature performance stems from the degradation of g_m and f_T. Therefore, the better temperature stability of the SiGe LNA is due to the much higher f_T over the whole temperature range. Furthermore, the bias network in Fig. 6.3 implements the aforementioned compensation of the finite β, partially also compensating for temperature variations. Even at the highest required temperature for automotive applications the performance of both LNAs would still be suitable for sufficient receiver performance. However, large variations of CMOS LNA may affect the sensitivity of the system over the wide temperature range.

6.1.4 LNA Results Summary and Comparison

The measurement results of both LNAs are summarized in Table 6.1 and compared with state-of-the-art designs. Both LNAs designed in this work offer a performance comparable with the state-of-the-art in terms of noise figure, gain, linearity, power consumption and area. They also offer high ESD protection. The SiGe LNA compares favorably to literature in terms of NF and linearity.

Compared to the CMOS LNA, the SiGe LNA offers better noise figure and linearity at the expense of higher power consumption. Additionally, the SiGe circuit offers better performance stability over a wide temperature range. The CMOS LNA offers slightly better gain, since it implements two stages, whilst consuming similar chip area. Both circuits offer comparable high ESD protection levels.

Table 6.1 LNA performance comparison

Feature	[22]	[23]	CMOS	SiGe	[24]	[16]
Technology	0.18 μm CMOS	0.1 μm SOI	0.13 μm CMOS	0.35 μm SiGe	0.17 μm SiGe	0.25 μm SiGe
f_T (GHz)	45	95	100	200	170	130
Topology	SE	SE	Diff	Diff	Diff	Diff
BW (GHz)	22-25[1]	21.8-25.8[1]	22.7-25	22.2-26	18-25[1]	8-26[1]
f_0 (GHz)	23.7	23.8	24	24	21[1]	24
Gain (dB)	12.9	7.3	14	12	22.5	10.5
NF(dB)	5.6	10	5	3.1	3.2	4.7
IP1dB (dBm)	-11.1	-16.2	-14.1	-8.7	-15.5	-18.5
IIP3 (dBm)	2.04	-7.8	-1.7	-1.8	NA	-6.0
S_{11} (dB)	-11	-45	-7	-5.4	-12	-19.1
S_{22} (dB)	-22	-9.4	-15	-10.6	-13	-26.3
TLP (A)	NA	NA	2	1.7	NA	NA
P_{dc} (mW)	54	79.5	18	41	42	40
V_{dc} (V)	1.8	1.5	1.5	3.3	1.2	3.3
Size (mm^2)	0.735	0.22	0.23	0.32	NA[2]	0.33

[1] Graphically estimated.

[2] LNA is part of a receiver. The area of the stand-alone LNA is not specified.

6.2 Mixers

Mixers are key components of radar receiver circuits since they realize frequency conversion. In a receiver they are used to down-convert an RF signal to an intermediate frequency (IF). As discussed in section 2.6, direct down-conversion receiver architectures are advantageous due to their simplicity compared to double-conversion or "sliding-IF" topologies. Thus, the mixers in this work are designed for direct or zero-IF down-conversion.

There are numerous possibilities for the circuit implementation of mixers. These are typically distinguished with respect to the used device and the allocation of the signals to the LO, RF and IF ports. Additionally, mixers can be classified with respect to DC bias into active and passive mixers. Active mixers are usually designed to provide conversion gain, whilst passive mixers typically have conversion loss, but offer good linearity and minimal DC power consumption. Active mixers, based on Gilbert cell topology [25], are highly popular both in CMOS and SiGe technologies. The implementation of passive mixers usually depends on the available non-linear devices in a given technology. In CMOS, the passive mixers are typically realized using MOSFETs as voltage-controlled resistors. These resistive mixers are very popular due to their simplicity and the possibility of overcoming the flicker noise problem of MOSFETs, inherent to Gilbert mixers. Alternatively, the non-linearity of diodes can be used to realize a passive mixer. Several silicon-based processes provide an option of a very fast Schottky diode suitable for passive mixers at microwave frequencies.

6.2 Mixers 93

Mixers can be also classified with respect to the circuit topology into single-ended, single-balanced and double-balanced [26]. Passive ring mixers belong to the category of double-balanced mixers. The double-balanced topology is preferable, since it offers a very good isolation between the ports, improved suppression of spurious products and differential circuit topology. This is particularly advantageous for narrowband systems due to minimal separation between the RF and LO tones. Both single-ended and double-balanced topologies have been designed in this work and described in this chapter.

Large amount of works has been published on down-conversion mixers at microwave and millimeter-wave frequencies. Most publications on active mixers focus on high linearity [27], low power consumption [28], low flicker noise [29] or low noise figure [30], whilst passive mixers are usually optimized for minimal conversion loss [31], very wide bandwidth [32] or highest linearity [33].

Section 6.2.1 describes and compares 24 GHz double-balanced active mixers in CMOS and SiGe realized in this work and presented in [34]. Section 6.2.2 describes and compares passive mixers in CMOS and SiGe technologies presented in [35] and [36], respectively. Finally, section 6.2.3 compares qualitatively the properties of active and passive mixers for receiver front-ends.

6.2.1 Active Mixers

6.2.1.1 Active Mixer in CMOS Technology

The conceptual schematic diagram of the double-balanced active CMOS mixer is presented in Fig. 6.13. The circuit is based on the Gilbert cell.

The transistors M1 and M2 compose the differential pair used for the RF transconductance stage. The design of a common-source low-noise amplifying stage in CMOS is very sensitive to transistor layout and series resistance of the matching inductors [37]. Therefore, in CMOS also the layout-dependent gate resistance R_g has to be taken into consideration. As can be seen in Eq. (3.1), there is an optimum transistor width for lowest R_g. Furthermore, as shown in Fig. 3.10(a), in C11N the minimal noise figure of a low-Vt NMOSFET is achieved around 0.1 mA/μm, which is considerably below the optimum current density for peak f_T and g_m. Thus, choosing the transistor width poses a trade-off between noise figure, gain and current consumption. Larger width would be advantageous for higher transconductance and thus higher gain, but at the cost of higher current consumption, increased noise figure and higher parasitic capacitance. In this design the amplifying stage has been optimized for lowest noise figure. The transistors have been dimensioned such that the optimum noise impedance is close to 50 Ω. The corresponding optimal gate width was found to be 55 μm. The parasitic capacitance at the input nodes is tuned out at 24 GHz by the series inductances L1, L2 at the gates of M1, M2.

The transistors M3-M6 are used as the LO switching stage. Following [38], the series peaking inductors L3 and L4 have been added between the transconductance

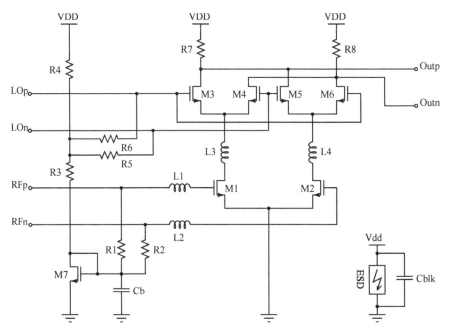

Fig. 6.13 Schematic diagram of the active CMOS mixer.

and switching stages in order to improve the conversion gain of the CMOS mixer by partially tuning out the parasitic capacitance at the drains of M1 and M2 and at the sources of M3-M6. It has been observed in simulation that implementation of these inductors contributes to the performance improvement to a higher extent rather than if they are attached at the sources of M1 and M2. Therefore, the classical source degeneration technique has been disregarded in this circuit due to area considerations.

The bias network of the RF transconductance stage consists of a diode-connected transistor M7, capacitor Cb and large resistors R1, R2. The large value resistors R1 and R2 are used to isolate the bias network from the RF signal. The current mirror between the transistor M7 to M1 and M2 sets the total current in transistors M1 and M2 to four times the reference branch current. The gate potential of the LO switching stage, applied through large resistors R5 and R6, is set by a resistive divider between R3 and R4. In order to improve the stability over process variations and tolerances, matching of the transistors M7 to M1, M2, resistors R2 to R1 and R4 to R5 has been carefully taken into consideration. Furthermore, large resistors R1 and R2 isolate the RF path from the DC bias network. The capacitor Cb provides a low-impedance path and thus virtual ground for high frequency signals.

The voltage conversion gain of a Gilbert mixer is given using the well-known expression [39]

$$A_v = \frac{2}{\pi} \cdot g_m \cdot Z_L, \tag{6.21}$$

where g_m is the transconductance of the RF stage and Z_L is the load impedance. Therefore, the gain can be increased by using a larger load impedance at the intermediate frequency. The loads in the circuit are realized as resistors, since the circuit has to operate down to zero-IF. The choice of the load resistors R7 and R8 poses a trade-off between gain and linearity. During the operation point optimization, particular attention has been paid to maximize the possible swing across the resistors R7 and R8, whilst still keeping the transistors of the RF transconductance stage in saturation. This is necessary to achieve optimal linearity. Thus, the optimal value for the load resistors R7-R8 was found to be 350 Ω.

The blocking capacitor Cblk has been added to provide a low-impedance path for high-frequency signals between VDD and GND. Additionally, an ESD power-clamp device has been added between the power supply and ground.

6.2.1.2 Active Mixer in SiGe Technology

The schematic diagram of the bipolar mixer is presented in Fig. 6.14. It has a double-balanced Gilbert cell topology similar to that of the active CMOS mixer.

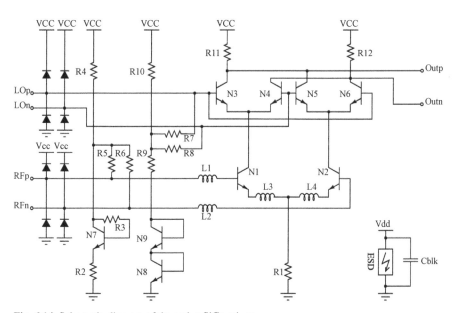

Fig. 6.14 Schematic diagram of the active SiGe mixer.

The design has been performed similarly as for the active CMOS mixer in the previous section. The transistors N1, N2 compose the differential pair used for the RF transconductance stage. The stage has been designed for low thermal noise figure according to the procedure presented in [7] and implemented for the SiGe LNA design in section 6.1.2.

The input impedance of a bipolar common-emitter stage has a much lower Q-factor compared to the input impedance of a MOSFET common-source stage. This is due to the presence of an equivalent input conductance g_π in a bipolar transistor, compared to a small resistive component introduced by the gate resistance R_g in a MOSFET [37]. Therefore, bipolar transistors are easier to match over wider bandwidth and are less sensitive to impedance mismatch. Thus, the design of a common-emitter stage in SiGe is more straightforward than a design of its equivalent in CMOS and classical matching techniques can be used. In this case, inductors L1, L2 at the bases and L3, L4 at the sources of transistors N1, N2 have been designed for simultaneous noise and impedance matching.

As mentioned previously, following [18] the optimal linearity is achieved for high C_{bc} and low r_b. Therefore, similarly as in the SiGe LNA, the transistors N1 and N2 have been laid-out in the double-base contact configuration CBEBEBC in order to reduce the base resistance and thus improve noise figure and linearity.

The transistors N3-N6 are used as the LO switching stage. The size of the transistors is chosen as a compromise between low base resistance r_b that requires large transistor size and fast switching speed that requires smallest parasitics and small size. The optimal transistor configuration was found to be BEBC, described in Fig. 3.6, with emitter length of 10 μm.

Similarly as described previously, during setting of the operating point the load resistors were chosen such that the IF output swing can be maximized for better linearity, whilst the transistors of the RF transconductance stage remain always in the active region. In order to have better gain, the mixer has been designed for high IF load impedance. The optimal load resistors R11-R12 in the SiGe circuit were found to be 130 Ω. Obviously, designing the bipolar mixer for 3.3 V provides an advantage over the CMOS mixer due to larger available headroom. This allows to achieve better linearity.

The bias network for the RF stage of the SiGe mixer is very similar to that of the CMOS mixer. It is based on a current mirror between N7 and N1, N2, whilst the ratio between the resistors R1 and R2 must be set accordingly. The total current in transistors N1 and N2 is four times the reference branch current. In order to improve the performance stability over process variations and tolerances, matching of the transistors N7 to N1,N2, resistors R2 to R1 and R3 to R4, R5 has been carefully taken into consideration. The resistor R1, attached at the virtual ground node of the differential pair consisting of N1 and N2, is also advantageous for improving the common-mode stability of the circuit.

The feedback around the transistor N7 provides compensation for finite β. When β is reduced, as a result e.g. of a temperature increase, the collector current of N7 is increased and thus higher current is mirrored to N1 and N2. The resistor R3 is essential for symmetry considerations of the current mirror between N7 and N1, N2, to compensate for the voltage drop on R5 and R6 due to the base current into N1, N2.

6.2 Mixers 97

For the LO switching stage the bias is considerably simpler. It consists of two diode-connected transistors N8 and N9 and a resistive divider between R9 and R10, which sets the potential at the bases of N3-N6. The bias is set through large value resistors R7 and R8 that isolate the bias network from the LO input.

Similarly as for the CMOS mixer, blocking capacitor Cblk and ESD power clamp have been added between VCC and GND. Additionally, due to margins of performance of the SiGe circuit, it was possible to attach ESD protection diodes directly at the RF pins, following the classical I/O protection principle described in section 6.6.1. The ESD diode devices have been optimized for lowest parasitic capacitance of around 60 fF. Thus, the introduced performance deterioration due to the directly attached ESD devices at 24 GHz is still negligible.

6.2.1.3 Measurements of CMOS and SiGe Active Mixers

The active mixer circuits, described in sections 6.2.1.1 and 6.2.1.2, have been processed in Infineon's C11N and B7HF200 technologies, respectively. The annotated chip micrographs of the CMOS and SiGe mixers are presented in Fig. 6.15. The size including the pads is 0.24 mm^2 for the CMOS and 0.32 mm^2 for the SiGe chip, respectively.

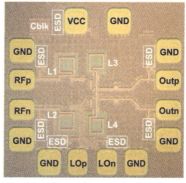

(a) CMOS (0.52 mm × 0.47 mm)   (b) SiGe (0.59 mm × 0.55 mm)

Fig. 6.15 Chip micrographs of the active mixer circuits.

The inductors are designed as spiral coils in order to minimize the chip area. Smaller area is also advantageous for minimizing on-chip wiring, thus having less losses spent on the on-chip interconnects. The windings of the coils are realized in both technologies in the top metallization layer in order to keep the inductors as away as possible from the substrate to minimize substrate losses and achieve the highest quality factor. The ground shielding technique [40] for improving the quality factor has been disregarded in this case, since it causes the self-resonance frequency (SRF) to drop significantly [41] due to an additional capacitance between the top metal and the ground shield. The undercrossings of the inductors have been realized in the layer below the top level. The spiral coils have been designed to exhibit the maximum quality factor at the center frequency of 24 GHz.

Similarly, as described in section 6.1.3, the CMOS chip was thinned to 185 μm and the SiGe chip to 350 μm in order to keep bondwires as short as possible. The thicknesses of the chips are different due to the fact that only certain thicknesses were available on C11N and B7HF200 shared-reticle runs. Both thinned chips have been mounted on a test PCB shown in Fig. 6.16 and fully characterized on-board. The board implements hybrid ring couplers for single-ended to differential conversion at the RF and LO ports. It can be noticed that high-frequency connectors have been also used at the IF ports. This is necessary in order to enable measurement of the LO and RF frequency components at the IF port during isolation characterization.

Fig. 6.16 Double-balanced mixer characterization board.

The measurements have been performed using Agilent's 8970B Noise-Figure-Meter (NFM), 8565EC Spectrum Analyzer (SPA) and 83650A synthesized sweepers. The measurement setups are described in detail in Appendix D.3. Similarly as for the measurement of LNAs, setup losses have been de-embedded using the *Insertion Loss* technique described in section 5.2 and the power levels were accurately determined using power sensors. Both mixers have been measured using a buffer that offers a high load impedance at the IF output.

The CMOS mixer consumes 2.8 mA from 1.5 V, whilst the SiGe mixer consumes 12 mA from a single 3.3 V supply. Both chips have been characterized for RF input frequencies from 21.5 GHz to 26.5 GHz, whilst the IF was kept constant at 10 MHz. The input LO power was 3 dBm and the input RF power −20 dBm. The measured and simulated conversion gain and noise figure versus RF frequency are presented in Fig. 6.17 for CMOS and in Fig. 6.18 for SiGe, respectively. As can be observed, both circuits achieve a very low DSB noise figure of 7.5 dB for CMOS and 4.7 dB for SiGe around the center frequency of 24 GHz. The bipolar

6.2 Mixers

circuit offers a higher gain of 10.5 dB compared to 7 dB of CMOS. As explained in section 5.2, the measured frequency characteristics are deformed compared to simulation due to de-embedding inaccuracy. However, in the narrow-band around the center frequency of 24 GHz a very good match between the measured and simulated results has been achieved due to accurate modeling of the parasitics.

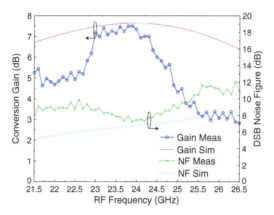

Fig. 6.17 Measured and simulated gain and noise figure of active CMOS mixer.

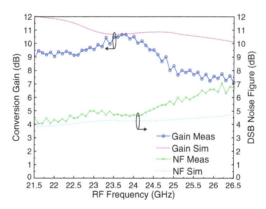

Fig. 6.18 Measured and simulated gain and noise figure of active SiGe mixer.

The input-referred 1dB compression-point was measured to be -12 dBm for CMOS and -7.2 dBm for SiGe at an RF frequency of 24.01 GHz and an IF frequency of 10 MHz. The IIP3 measurement has been performed by applying 24 GHz at the LO port and two input tones 24.01 GHz and 24.011 GHz at the RF port and observing the down-converted harmonics at 9, 10, 11 and 12 MHz. The CMOS and SiGe circuits exhibit IIP3 of -1.4 dBm and -0.8 dBm, respectively. The IP1dB and

IIP3 linearity measurements of CMOS and SiGe mixers for an LO power of 3 dBm are presented in Fig. 6.19 and 6.20, respectively.

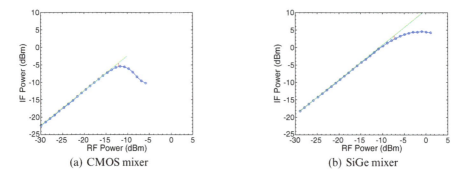

Fig. 6.19 IP1dB measurement of the active mixers in CMOS and SiGe.

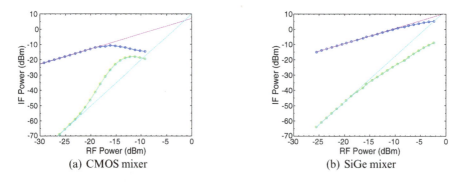

Fig. 6.20 IIP3 measurement of the active mixers in CMOS and SiGe.

The isolation between the ports LO-RF, LO-IF and RF-IF over the input frequency for CMOS and SiGe mixers is presented in Fig. 6.21. As can be seen, both mixers achieve good isolation above 26 dB over the whole frequency range.

The robustness of the mixers has been verified over a wide range of temperatures from $-40\,°C$ to $125\,°C$. Fig. 6.22(a) presents gain and noise figure as a function of temperature for the CMOS and Fig. 6.22(b) for the SiGe circuit, respectively. As expected, in both cases the performance at lower temperatures is improved. For the CMOS mixer the gain drops by 4 dB and noise figure increases by nearly 5 dB between the lowest and the highest temperature. For the SiGe mixer the gain is much more stable and drops only by 1.7 dB over the whole temperature range, whilst the noise figure increases by 3 dB. Similar to the description in section 6.1.3, also here the stronger performance deterioration of the CMOS circuit at high temperature is

6.2 Mixers   101

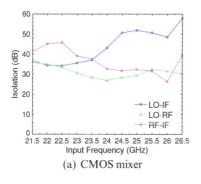
(a) CMOS mixer

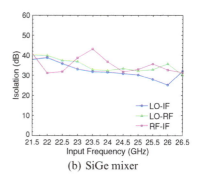
(b) SiGe mixer

Fig. 6.21 Measured port to port isolation active mixers in CMOS and SiGe.

due to lower f_T of NMOSFET compared to HBT *npn* device. The circuits are biased for optimum noise figure, therefore no proportional to absolute temperature (PTAT) network was used.

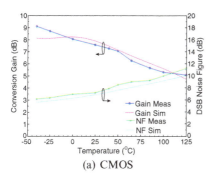
(a) CMOS

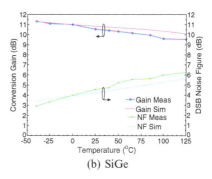
(b) SiGe

Fig. 6.22 Measured gain and noise figure of active mixers versus temperature.

6.2.1.4 Active Mixers Results Summary and Comparison

The measurement results of both active mixers are summarized in Table 6.2 and compared with the state-of-the-art differential designs. Both circuits compare favorably to literature in terms of noise figure and power consumption in case of the CMOS mixer and in terms of gain, noise figure, linearity and chip size in case of the SiGe mixer. In comparison to CMOS, the SiGe mixer provides better gain, noise figure and linearity at the expense of higher power consumption. Additionally, the SiGe circuit offers better performance stability over a wide temperature range and ESD protection on the RF pins. However, it has to be noted that the CMOS mixer has a much lower tail current than the SiGe mixer.

Table 6.2 Active mixer performance comparison

Feature	[42]	[43]	CMOS	SiGe	[24]	[28]
Technology	90 nm SOI CMOS	0.13 μm CMOS	0.13 μm CMOS	0.35 μm SiGe	0.17 μm SiGe	0.18 μm SiGe
f_T (GHz)	160	76	100	200	170	140
f_{RF}/f_{IF} (GHz)	30/2.5	30/0.01	24/0.01	24/0.01	24/0.1	24/1
Gain (dB)	-2.6	7.5	7	10.5	9.5	4.8
DSB NF(dB)	10.5	16.4	7.5	4.7	4.8	10.2
IP1dB (dBm)	NA	NA	-12	-7.2	-18	-32
IIP3 (dBm)	1.8	1.2	-1.4	-0.8	NA	-16.6
LO Power (dBm)	5	5	3	3	0	-6.5
LO-IF (dB)	NA	70	42.9	31.4	NA	NA
RF-IF (dB)	NA	55	32.4	36.5	NA	NA
LO-RF (dB)	NA	31	26.6	32.2	47	NA
P_{dc} (mW)	20	97	4.2	39	37	6.4
V_{dc} (V)	1.2	3.3	1.5	3.3	2.5	1.0
Size (mm^2)	0.2	0.25	0.24	0.32	NA	1.31

6.2.2 Passive Mixers

6.2.2.1 Passive Resistive Ring Mixer in CMOS Technology

Passive resistive mixers are based on field-effect transistors (FET) acting as switches. A large LO signal, applied to the gate, modulates the channel resistance of the FET. An RF signal, applied to the drain or source, is multiplied with a square-wave-like time-varying function, having a fundamental frequency component at the LO frequency. The resulting mixing product at the difference frequency, filtered from the drain or source, is used as the IF signal in down-conversion mixers. Even though the gate terminal may be DC biased at a certain potential, the resistive mixer is a passive one, since its drain-source voltage is DC biased at 0 V to keep the transistor in the deep-triode region.

The passive resistive mixer can be realized in doubled-balanced configuration using the ring topology. The schematic diagram of the passive resistive ring mixer is presented in Fig. 6.23.

In order to reduce the conversion loss for moderate LO power, the gates of the mixing transistors M1-M4 are biased in the vicinity of the threshold voltage. It has been found in simulation that the optimal biasing potential in this case is 400 mV. Thus, a bias network has been integrated on-chip to set the potential at the gates of the mixing transistors. The bias voltage is generated by a small current of 0.2 mA through the diode-connected transistor Mb. The resistors Rb2 and Rb3 of 5 kΩ are used to isolate the LO input from the bias network and apply the potential to the gates. A bypass capacitor Cb of 2 pF is attached at the gate of Mb to stabilize the voltage. It has been observed in measurement that applying the bias potential re-

6.2 Mixers

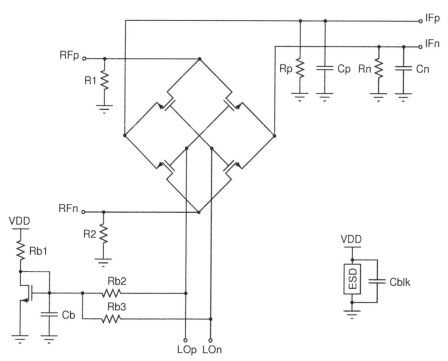

Fig. 6.23 Schematic diagram of the passive resistive CMOS mixer.

duces the loss by as much as 3 dB for a moderate LO power up to 6 dBm, whilst for high LO power it does not have effect.

The overall conversion loss of a resistive mixer can be simplified to the form

$$L_{\text{total}} = L_{\text{idler}} + L_{\text{mix}} + L_{\text{res}}, \tag{6.22}$$

where L_{idler} is due to the power lost in unwanted higher order mixing products (idler frequencies), L_{mix} is due to the time varying drain-source resistance under LO pumping, deviating from the optimum resistance waveform derived in [44] and L_{res} is due to non-ideal resistance values of the on and off states.

The component L_{idler} is minimized, by providing filters offering optimal terminations at the idler frequencies. However, there is a theoretical lower bound on conversion loss of 3.92 dB due to the energy lost in unwanted mixing products [45]. The loss L_{mix} is reduced by making the drain-source resistance waveform closer to rectangular and optimizing its duty cycle [44]. Finally, L_{res} is related to the transistor parameters. The minimal conversion loss due to non-ideal resistance values is given by [46]

$$L_{\text{res}} = 1 + 2\gamma^2 \left[1 + \sqrt{1 + \left(\frac{1}{\gamma}\right)^2} \right], \tag{6.23}$$

where

$$\gamma = \left| \frac{Z_{\text{on}}}{Z_{\text{off}}} \right|, \tag{6.24}$$

where Z_{on} and Z_{off} are the on- and off-resistances, respectively. The expressions for these resistances can be easily derived from an equivalent circuit representation looking into the RF port of a resistive FET mixer [5]

$$Z_{\text{on}} \approx R_s + R_d + \frac{L_g}{\sigma W_g} \approx \frac{L_g}{\sigma W_g}, \tag{6.25}$$

$$Z_{\text{off}} \approx R_d + \frac{1}{j\omega C_{\text{gd}}} \approx \frac{1}{j\omega L_g W_g C_{\text{gd},a}}, \tag{6.26}$$

where the source and drain parasitic resistance contributions R_s and R_d are neglected, L_g is the gate length, σ is the channel conductivity, W_g is the gate width, C_{gd} is the drain-source capacitance and $C_{\text{gd},a}$ is the capacitance per area. As can be seen in Eq. (6.23), the loss component L_{res} is minimized by reducing γ. However, plugging Eq. (6.25) and (6.26) into (6.24) shows that γ is only related to technology dependent parameters, whilst the gate length L_g is set to the minimum allowed by the given technology. Nevertheless, the gate width W_g can be optimized in the design as a trade-off between linearity, port to port isolation, requirement for lower LO power and matching. In this case an NMOSFET width of 60 μm has been found to be a good compromise for the mixing transistors.

The option of having inductors to build a shunt LC filter at the IF frequency, as proposed in [47] has been disregarded, due to the fact that this mixer is required to function down to very low IF frequencies. This would make the realization of the inductors impractically large and inefficient. Thus, the DC path for the IF side at the source of the mixing transistors has been realized by attaching large value resistors Rp, Rn of 10 kΩ.

Furthermore, the option of an RF filter has been disregarded, due to the fact that RF and LO frequencies are very close and impractically sharp high-Q filters would be required. Therefore, the isolation shall be achieved by using the differential ring structure. The DC path for the RF side is provided by attaching large resistors R1, R2 of 10 kΩ at the drains of the mixing transistors.

In order to attenuate the high frequency content at the IF output, capacitors Cp and Cn of 4 pF have been attached. This provides sufficiently low resistance path to ground for the RF and LO signals.

Similarly as in the active mixer, a power clamp ESD device and the bypass capacitor Cblk have been attached between VDD and GND.

The presented inductorless passive resistive mixer approach allows to achieve wide bandwidth and lowest chip area. A similar passive resistive mixer using single-ended topology is presented in in [48]. It achieves $0.5 - 25$ GHz bandwidth and consumes a minimal active area of 0.014 mm^2. The matching of that mixer is similarly optimized by the size of the transistors, whilst the conversion loss is minimized by integrating an on-chip bias network.

6.2.2.2 Passive Bipolar Mixer in SiGe Technology

The silicon-based SiGe:C bipolar technology offers the integration capability along with excellent high-frequency characteristics. However, unlike the MOS transistor, the HBT transistor is less suitable as a voltage switch for the realization of a passive mixer. Thus, the passive resistive topology described in the previous section is not applicable. An alternative very common option to realize passive mixers is to use diodes.

Several silicon-based processes provide an option of a very fast Schottky diode with cut-off frequencies up to terahertz [49]. Schottky diodes are particularly suitable for implementation of passive mixers at microwave and millimeter-wave frequencies. However, in several processes, as e.g. in B7HF200 SiGe:C bipolar technology, this option is not available. Furthermore, using an additional process option might be disadvantageous due to cost considerations. Thus, suitability of diode-connected npn transistors for realization of an integrated passive diode mixer has been investigated in this work.

Balanced mixer topology is preferred due to the superior isolation. Thus, passive diode mixers are typically implemented using either a transformer or a coupler to split the input LO or RF signal and to provide it in anti-phase to the mixing diodes. However, it is a challenge to realize a hybrid coupler on-chip at microwave frequencies due to large area consumption. Therefore, such mixers are usually integrated as discrete components on board [50] or ceramic substrate [51], whilst at millimeter wave frequencies the coupler is typically realized using transmission lines [52].

A compact hybrid ring coupler has been realized in this work on-chip using lumped elements and described in detail in section 6.5.1.3. Therefore, a compact on-chip realization of a single-balanced passive diode mixer became possible. The diagram of the realized direct down-conversion mixer is presented in Fig. 6.24. It is based on a classical topology described in [17]. The RF and LO signals are applied at the Δ and Σ or vice versa at Σ and Δ ports of a hybrid ring coupler, respectively. These ports are mutually isolated, thus ensuring sufficient LO-RF isolation. A hybrid ring coupler provides the RF signal in anti-phase and LO signal in-phase to the diodes. Furthermore, at the IF port a low impedance path to ground is required for RF and LO signals. This is achieved using a filter that is also required to offer termination at the IF frequency.

The hybrid ring coupler in Fig. 6.24 can be implemented using lumped resonant circuits. The $\lambda/4$ and $3/4\lambda$ lines are substituted by lumped element equivalents that generate the required phase conditions at the center frequency [39]. The schematics of the mixer circuit, including the lumped element hybrid ring coupler realization, is presented in Fig. 6.25.

The components of the hybrid ring coupler are easily dimensioned by [53]

$$
\begin{aligned}
L &= \frac{Z_0 \cdot \sqrt{2}}{\omega_0}, \\
C &= \frac{1}{Z_0 \cdot \sqrt{2} \cdot \omega_0}.
\end{aligned}
\tag{6.27}
$$

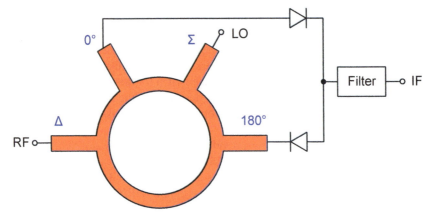

Fig. 6.24 Diagram of a classical single-balanced diode mixer.

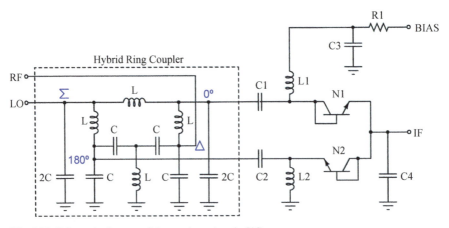

Fig. 6.25 Schematic diagram of the passive mixer in SiGe.

where Z_0 is the port impedance of the coupler, which is 50 Ω here, and $\omega_0 = 2\pi f_0$ is the center angular frequency, where f_0 is the center frequency of 24 GHz in this case.

The *npn* transistors N1, N2 with shorted base-collector terminals are used to realize the diodes. The continuous DC path through the diodes, required for a single-balanced mixer [26], is provided through the inductors L1, L2. In order to minimize the conversion loss, a bias potential is applied at the collector of N1. The diode-connected transistors are forward-biased slightly above the knee voltage of 0.7 V. The bias current is limited by the large resistor R1 of 1 kΩ, whilst C3 is used for stabilizing and cleaning the bias voltage.

The capacitors C1, C2 along with inductors L1, L2 are used for matching between the coupler and the diode-connected transistors N1, N2. Furthermore, the

6.2 Mixers

capacitors C1, C2 are also used for AC coupling in order to avoid short between the bias potential and ground through the coupler. The inductors L1, L2 are used to short the IF component at the high-frequency side.

At the IF output of the mixer, the capacitor C4 is used to provide a low-impedance path to ground for high-frequency LO and RF components.

6.2.2.3 Measurements of CMOS and SiGe Passive Mixers

The passive mixer circuits, described in sections 6.2.2.1 and 6.2.2.2, have been processed in Infineon's C11N and B7HF200 technologies, respectively. The annotated chip micrographs of the CMOS and SiGe mixers are presented in Fig. 6.26. The die size including the pads of the passive resistive ring mixer in CMOS is 0.23 mm^2, whilst the active mixer area is only 0.022 mm^2. The chip size including the pads of the passive SiGe mixer is 0.33 mm^2.

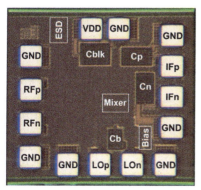
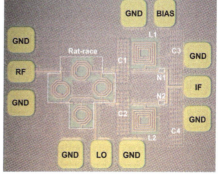

(a) CMOS (0.5 mm × 0.47 mm)   (b) SiGe (0.64 mm × 0.52 mm)

Fig. 6.26 Chip micrographs of the passive mixer circuits.

The CMOS chip was thinned to 185 μm and fully characterized on the board shown in Fig. 6.16. The SiGe chip has been characterized on-wafer using Cascade Microtech Infinity probes up to 67 GHz with 100 μm pitch in GSG configuration. Both mixers have been measured directly in a 50 Ω environment.

The CMOS mixer draws 0.2 mA from 1.5 V for the integrated bias network, whilst the SiGe mixer draws 0.8 mA from 1.5 V on the bias pad.

The CMOS and SiGe chips have been characterized over the RF input frequency for a constant IF frequency of 10 MHz and 100 MHz, respectively. The CMOS circuit has been measured in the frequency range from 19 GHz to 26.5 GHz, whilst the SiGe circuit has been measured from 20 GHz to 40 GHz. It has to be emphasized that frequency characterization of the double-balanced CMOS mixer was limited by the bandwidth of the on-board couplers. The input RF power was −20 dBm, whilst the input LO power was 6 dBm for CMOS and 3 dBm for SiGe mixer. The measured and simulated conversion loss versus RF frequency are presented in Fig. 6.27(a) for CMOS and in Fig. 6.27(b) for SiGe, respectively.

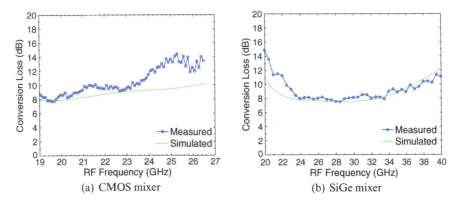

Fig. 6.27 Measured and simulated conversion loss of the passive mixers.

A conversion loss of 11.4 dB and 8 dB can be observed at 24 GHz for CMOS and SiGe circuits, respectively. Higher deviation of the conversion loss from the simulation at some frequency points above 24 GHz in Fig. 6.27(a) are due to the fact that the available sweeper could not provide sufficient LO power to drive the mixer at 6 dBm. A very wide bandwidth has been observed for the SiGe mixer. This can be easily explained by observing the on-wafer measurements of the ring coupler, described in detail in section 6.5.2.3. The structure offers a phase difference close to 180° with maximal amplitude imbalance of 2 dB over a frequency range from 20 – 40 GHz. A good anti-phase excitation of the mixing diodes provides the circuit functionality over the whole frequency range.

The input-referred 1dB compression-point was measured to be 5.1 dBm for CMOS and −1.5 dBm for SiGe. The LO frequency was 24 GHz and the IF frequency was 10 MHz for CMOS and 100 MHz for SiGe. The CMOS and SiGe circuits exhibit IIP3 of 11.8 dBm and 8.4 dBm, respectively. The IIP3 was measured by applying two tones similarly to the description in section 6.2.1.3. The IP1dB linearity measurements of CMOS and SiGe mixer circuits for LO power of 6 dBm and 3 dBm are presented in Fig. 6.28(a) and 6.28(b), respectively.

An additional important aspect of passive mixers is the minimal LO power, required in order to achieve optimal performance. It has been observed from measurements of conversion losses versus LO input power, presented in Fig. 6.29, that moderate power levels of 6 dBm and 3 dBm are sufficient for CMOS and SiGe mixers, respectively.

The isolation between the ports LO-RF, LO-IF and RF-IF over the input frequency for CMOS and SiGe mixers is presented in Fig. 6.30. As can be observed, both mixers achieve good isolation over the whole frequency range. For the CMOS mixer the isolation is higher than 28 dB over the whole range, whilst for the SiGe

6.2 Mixers

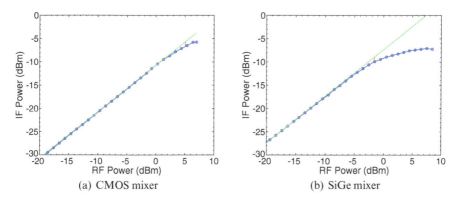

Fig. 6.28 IP1dB measurement of the passive mixers in CMOS and SiGe.

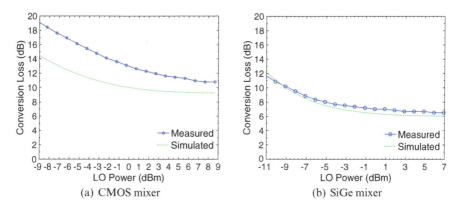

Fig. 6.29 Measured and simulated loss versus LO power of the passive mixers.

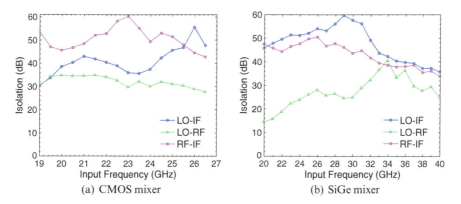

Fig. 6.30 Measured port isolation of passive mixers in CMOS and SiGe.

mixer a lowest LO-RF isolation of 14 dB has been observed at 20 GHz. The LO-RF isolation is determined by the frequency characteristics of the ring coupler.

6.2.2.4 Passive Mixers Results Summary and Comparison

The measurement results of both passive mixers are summarized in Table 6.3 and compared with the published works. The CMOS mixer compares favorably to literature in terms of isolation, required LO power and area, whilst the SiGe mixer offers better conversion loss and requires lower LO power. However, the comparison of the SiGe mixer with the state-of-the-art is not straightforward, since according to the author's knowledge this is the first reported mixer integrated in SiGe technology without a Schottky diode.

Comparing the two presented passive mixers, the single-balanced diode mixer in SiGe provides better conversion loss, higher port LO-IF port isolation, wider bandwidth and requires less LO power. However, the passive resistive ring in CMOS offers better linearity and higher LO-RF isolation. Additionally, it has to be noted that the proposed inductorless approach for the passive mixer in CMOS has much smaller active circuit area compared to the bipolar mixer.

Table 6.3 Passive mixer performance comparison

Feature	[32]	[47]	CMOS	SiGe	[54]	[52]
Technology	0.13 μm CMOS	90 nm SOI CMOS	0.13 μm CMOS	0.35 μm SiGe	0.17 μm pHEMT	0.18 μm pHEMT
f_T (GHz)	NA	160	100	200	NA	381
Topology	Ring	SE[1]	Ring	SB[2]	SB[2]	SB[2]
BW (GHz)	1-11	26.5-30	19-26.5	22-39	46-78	NA
f_{RF}/f_{IF} (GHz)	6/0.5	27/2.5	24/0.01	24/0.1	46/1	76.5/0.1
Loss (dB)	7	10.3	11.4[1]	8	10	9.5
SSB NF(dB)	NA	11.4	13.2	NA	NA	NA
IP1dB (dBm)	5	NA	5.1	-1.5	NA	NA
IIP3 (dBm)	9	12.7	11.8	8.4	NA	NA
LO Power (dBm)	9	0	6	3	12.5	7
LO-IF (dB)	NA	22	37.5	51	NA	NA
RF-IF (dB)	NA	33	49	48	NA	NA
LO-RF (dB)	37	24	30	24	30	30
Size (mm^2)	0.62	0.12	0.23	0.33	1.5	0.29

[1] Single-ended (SE) resistive mixer.
[2] Single-balanced (SB) diode mixer.

6.2.3 Comparison of Active and Passive Mixers

Active mixers provide gain and can be optimized for a very low noise figure in the thermal noise frequency region. The presented active mixers in CMOS and SiGe exhibit, according to the author's knowledge, the lowest reported DSB noise figures of 7.5 and 4.7 dB, respectively. However, for an active CMOS mixer operating at very low intermediate frequencies down to few hertz, the flicker noise poses a major challenge. Several circuit techniques have been proposed in order to minimize the flicker noise in active mixers. An additional common approach is to use passive resistive mixers that can offer considerably lower flicker noise due to lack of the bias direct current.

Passive mixers have the advantage of minimum power consumption, very high linearity and simplicity. The main disadvantage is that they have conversion loss that has to be compensated in the following low-noise amplifier stage in the baseband. Conversion loss values of 11.4 and 8 dB have been obtained for the presented designs. Furthermore, passive mixers require higher LO power. In case that several passive mixers have to operate simultaneously from a single VCO, a very high LO power is required, which might not be available.

In addition to the presented measurements, also the flicker noise of the active double-balanced mixer, described in section 6.2.1.1, has been compared in measurement to that of the passive resistive ring mixer, described in section 6.2.2.1. As can be seen in Fig. 6.31, the passive mixer offers much lower flicker noise.

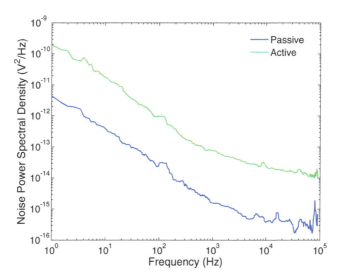

Fig. 6.31 Flicker noise comparison of active and passive mixers in CMOS.

Whether for a particular application active or passive mixers are more suitable, depends mainly on system requirements of the receiver. Thus, in order to compare active and passive mixers, the circuits have to be further integrated with other components. A detailed systematic analysis of system considerations on example of active and passive CMOS single-channel receivers consisting of an LNA and mixer is presented in the following section.

6.3 Single-Channel Receivers

The previous sections describe LNAs and mixers that have been developed in this work. The mixers have been designed as active and passive types and both options are compared in section 6.2.3. The question whether to implement an active or a passive mixer is posed for CMOS direct down-conversion receiver circuits, since in bipolar technology predominantly active mixers are used. The major motivation for using passive mixers in CMOS is the considerably lower flicker noise that they offer. However, as mentioned previously, in order to fully compare active and passive mixers, they have to be considered as part of a receiver. Only this allows to estimate and consider the system relevant aspects of implementing either option. Therefore, in the next step the LNA, active and passive mixer circuits presented in sections 6.1.1, 6.2.1.1 and 6.2.2.1, respectively, are integrated into two single-channel receivers in CMOS: LNA with active mixer and LNA with passive mixer.

In spite of the fact that differential signaling requires higher DC power consumption, it is advantageous in microwave circuits due to the superior noise and ground bounce immunity, better spurious response, decreased second-order non-linearity and improved stability. Therefore, the receivers implemented in this work use differential signaling and integrate only the double-balanced mixers.

Numerous publications have reported 24 GHz single-channel receiver circuits in CMOS using active mixers [55]. Some works describe receivers comprising LNA and passive mixers [31]. Only a few works in the literature compare active and passive down-conversion mixers. For instance, [56] compares double-balanced active and resistive ring stand-alone mixers at 1.8 GHz at room temperature, whilst [57] compares in simulation a double-balanced Gilbert cell with a resistive ring mixer for WCDMA applications at 2.1 GHz, particularly focussing on statistical simulations of IIP2. However, to author's knowledge, the first systematic comparison of active and passive receivers is presented in [58] and described in this section.

Section 6.3.1 presents the circuit design of both receivers. Section 6.3.2 describes the measurements of the two chips and analyzes the key receiver parameters such as gain, noise figure, power consumption, linearity, isolation and temperature stability. Finally, section 6.3.3 summarizes the results and draws conclusions from the comparison of receivers having active and passive mixers.

6.3.1 Design of Active and Passive Receivers in CMOS

The block diagrams of the receivers are presented in Fig. 6.32. Both receivers consist of an LNA and a mixer, connected by AC coupling capacitors. The receivers differ in the mixer stage. The first receiver, further referred to as *active*, uses the Gilbert mixer described in section 6.2.1.1, whilst the second receiver, further referred to as *passive*, uses the resistive ring mixer described in 6.2.2.1.

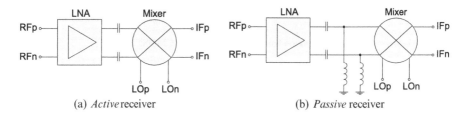

Fig. 6.32 *Active* and *passive* receiver block diagrams.

The LNA provides ESD protection at the RF port in both receivers using the virtual ground principle, as shown in Fig. 6.1. The active mixer additionally provides ESD structures attached directly at the LO and IF ports. The schematics of all the building blocks correspond to the described in the previous sections.

The active mixer achieves good matching to the differential impedance of 100 Ω on the RF inputs at the center frequency. Thus, no additional interstage matching was required in the *active* receiver. In the case of the *passive* receiver an interstage matching network has been added between LNA output and passive mixer input, consisting of series capacitors and shunt inductors, as shown in Fig. 6.32(b).

In order to achieve higher f_T and g_m, the gate length of the transistors used in both circuits was set to the minimal allowed gate length of the C11N technology. In both circuits blocking capacitors and ESD power clamp devices have been attached between VDD and GND.

6.3.2 Receiver Measurements and Analysis

Both receivers have been realized in Infineon's C11N technology. The annotated micrographs of the chips are presented in Fig. 6.33.

As described in previous sections, the chips were thinned to 185 μm in order to keep the bondwires short. The dies have been mounted and characterized on the board that was already used for the mixer and presented in Fig. 6.16. The receivers have been accurately characterized using a buffer that offers a high load impedance at the IF output, as described in Appendix D.3.

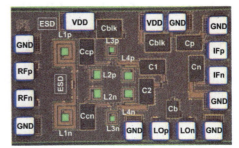

(a) *Active* receiver (0.86 mm × 0.59 mm) (b) *Passive* receiver (0.79 mm × 0.47 mm)

Fig. 6.33 Micrographs of the single-channel receiver chips in CMOS.

6.3.2.1 Chip Size

The die size including pads is 0.5 mm^2 and 0.37 mm^2 for the *active* and *passive* receiver chips in Fig. 6.33, respectively. The active area of the passive ring mixer of 0.022 mm^2 is much smaller than that of the active Gilbert mixer of 0.11 mm^2, but in both cases the die size is limited by the pads and the size of the LNA. Thus, the *passive* receiver has only slightly lower area than the *active* one. In case of a higher integration level of several receiver channels with additional RF, analog and digital blocks, the implementation of passive mixers could offer the advantage of a lower chip area.

6.3.2.2 Power Consumption, Gain and Noise Figure

The *active* receiver consumes 14.8 mA, whilst the *passive* receiver consumes 12.2 mA from a single 1.5 V supply. The difference is minor since the active mixer also has a low consumption of 2.8 mA.

The measured and simulated conversion gain and noise figure versus RF frequency of the *active* and *passive* receivers are presented in Fig. 6.34 and Fig. 6.35, respectively. Both chips have been characterized for RF input frequencies from 21.5 GHz to 26.5 GHz, whilst the the IF was kept constant at 10 MHz. The input LO power for the *active* receiver was 3 dBm and for the *passive* receiver 6 dBm. The input RF power was −20 dBm.

At the center frequency the *passive* receiver has a much lower gain of 2 dB compared to the 16 dB achieved in the *active* receiver. Therefore, an additional baseband amplifier is required to achieve the same gain of the front-end. This would require additional current consumption of up to 3 mA. Thus, under fair comparison conditions the *passive* receiver might not longer have the advantage of lower power consumption. Furthermore, the area of such an amplifier can be significant, if low flicker noise should be achieved. Very large transistors are required to achieve lowest flicker noise, as can be observed from the expression of the power spectral

6.3 Single-Channel Receivers

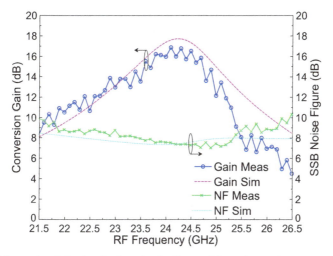

Fig. 6.34 Measured and simulated gain and noise figure of the *active* receiver.

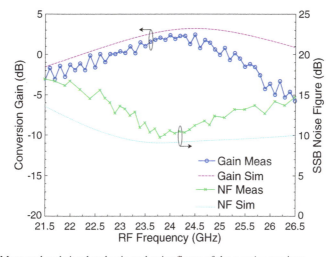

Fig. 6.35 Measured and simulated gain and noise figure of the *passive* receiver.

density

$$S_{\text{vf}} = \frac{K(V_{\text{gs}})}{C_{\text{ox}}} \frac{1}{WL} \frac{1}{f}, \quad (6.28)$$

of an equivalent input noise voltage source S_{vf}, modeling the flicker noise of a MOSFET [59], where S_{vf} has units $[\text{V}^2/\text{Hz}]$, $K(V_{\text{gs}})$ is a bias and process-dependent quantity, C_{ox} is the gate-oxide capacitance per unit area, W is the gate width, L is the gate length and f is the frequency. Thus, low-noise low-frequency amplifiers consume large chip area and the chip area advantage of passive mixers ceases.

116　　　　　　　　　　　　　　　　　　　　　　　　　　　　6 Radar Receiver Circuits

The total noise factor of a down-converting receiver, excluding the following stages, can be calculated by the Friis formula as follows [60]

$$F_{\text{Rec}} = F_{\text{LNA}} + \frac{F_{\text{Mix}} - 1}{G_{\text{LNA}}}, \tag{6.29}$$

where F_{LNA}, F_{Mix} and F_{Rec} are the single-side-band (SSB) noise factors of the LNA, mixer and receiver, respectively and G_{LNA} is the available power gain of the LNA. The receiver noise figure is then simply defined by expressing F_{Rec} in decibel, NF_{Rec} (dB) $= 10 \log F$. It can be observed from (6.29) that for sufficiently high LNA gain, the receiver NF is dominated by the NF of the LNA. In this case the *passive* receiver is expected to have a 1.2 dB higher NF than the *active* one due to much higher NF of the ring mixer compared to the Gilbert mixer. The measured SSB noise figure for the *active* and *passive* receivers at the center frequency are 8 dB and 10.5 dB. The larger difference is due to imperfect input and interstage matching.

The considerations made above are valid only for IF frequencies above the flicker noise corner, where noise is dominated by the thermal noise. For low-IF frequencies the noise of a receiver is dominated by the flicker noise of the mixers. Low frequency noise of the passive mixer is expected to be much lower compared to the active mixer. This can be also observed from measurement, presented in Fig. 6.31. Flicker noise corner frequencies, defined as the frequencies at which the total noise power is 3 dB higher than the thermal noise floor, of 1 MHz and 5 kHz and spectral power noise densities of $1.7 \cdot 10^{-11}$ V^2/Hz and $3.8 \cdot 10^{-13}$ V^2/Hz at an IF of 10 Hz have been obtained for the active and passive mixers, respectively.

6.3.2.3 Linearity

The input-referred 1dB compression point of both receivers, is given using the well-known chain formula

$$\frac{1}{\text{IP1dB}_{\text{Rec}} \, (\text{mW})} = \frac{1}{\text{IP1dB}_{\text{LNA}} \, (\text{mW})} + \frac{G_{\text{LNA}}}{\text{IP1dB}_{\text{Mix}} \, (\text{mW})}, \tag{6.30}$$

where $\text{IP1dB}_{\text{LNA}}$, $\text{IP1dB}_{\text{Mix}}$ and $\text{IP1dB}_{\text{Rec}}$ are the 1dB compression points of the LNA, mixer and receiver, respectively, and G_{LNA} is the available power gain of the LNA. To obtain a better insight, Eq. (6.30) can be rewritten as follows

$$\begin{aligned} \text{IP1dB}_{\text{Rec}} \, (\text{dBm}) = {} & \text{IP1dB}_{\text{LNA}} \, (\text{dBm}) + \text{IP1dB}_{\text{Mix}} \, (\text{dBm}) \\ & - 10 \log \left(\text{IP1dB}_{\text{Mix}} \, (\text{mW}) + \text{IP1dB}_{\text{LNA}} \, (\text{mW}) \cdot G_{\text{LNA}} \right). \end{aligned} \tag{6.31}$$

For an *active* receiver usually $\text{IP1dB}_{\text{LNA}} \, (\text{dBm}) + G_{\text{LNA}} \, (\text{dB}) > \text{IP1dB}_{\text{Mix}} \, (\text{dBm})$, which can be written in linear units as $\text{IP1dB}_{\text{LNA}} \, (\text{mW}) \cdot G_{\text{LNA}} > \text{IP1dB}_{\text{Mix}} \, (\text{mW})$. Thus, neglecting the term $\text{IP1dB}_{\text{Mix}}$ in the parentheses, Eq. (6.31) simplifies to

$$\text{IP1dB}_{\text{Rec}} \, (\text{dBm}) \approx \text{IP1dB}_{\text{Mix}} \, (\text{dBm}) - G_{\text{LNA}} \, (\text{dB}). \tag{6.32}$$

6.3 Single-Channel Receivers

This is also very intuitive, since the maximum allowable input power at the LNA input before the mixer runs into compression is the compression point of the mixer minus the gain of the LNA.

For a *passive* receiver usually $\text{IP1dB}_{\text{Mix}}\,(\text{dBm}) > \text{IP1dB}_{\text{LNA}}\,(\text{dBm}) + G_{\text{LNA}}\,(\text{dB})$, thus Eq. (6.31) simplifies to

$$\text{IP1dB}_{\text{Rec}}\,(\text{dBm}) \approx \text{IP1dB}_{\text{LNA}}\,(\text{dBm}). \tag{6.33}$$

Therefore, in this case only the linearity of the LNA is decisive for the overall linearity of the receiver.

Similar considerations are also applicable for the input-referred third-order intercept point, since the chain formula for the cascaded connection of receiver blocks [61] has the same form as in Eq. (6.30)

$$\frac{1}{\text{IIP3}_{\text{Rec}}\,(\text{mW})} = \frac{1}{\text{IIP3}_{\text{LNA}}\,(\text{mW})} + \frac{G_{\text{LNA}}}{\text{IIP3}_{\text{Mix}}\,(\text{mW})}. \tag{6.34}$$

The measured input-referred 1dB compression points for an RF frequency of 24.01 GHz are -24.6 dBm for the *active* and -13.4 dBm for the *passive* receiver, as presented in Fig. 6.36. The IIP3 for input tones at 24.01 GHz and 24.011 GHz have been measured to be -15.8 dBm and -4 dBm, respectively. As expected, the *passive* receiver offers better input-referred linearity, due to much lower gain.

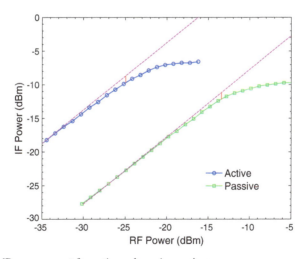

Fig. 6.36 IP1dB measurement for *active* and *passive* receiver.

However, for a fair comparison also the output-referred 1dB compression point (OP1dB) of the receiver, defined as [61]

$$\text{OP1dB}_{\text{Rec}}\,(\text{dB}) = \text{IP1dB}_{\text{Rec}}\,(\text{dB}) + G_{\text{Rec}}\,(\text{dB}) - 1, \tag{6.35}$$

where G_{Rec} is the total conversion gain of the receiver, should be considered. The *passive* receiver reaches -12.4 dBm, whilst the *active* receiver reaches a higher output power of -9.6 dBm in the linear operation region. Thus, the *active* receiver offers a higher output-referred 1dB compression point.

Additionally, a key system parameter that has to be considered is the intermodulation free dynamic range (IMFDR), also commonly referred to as the spurious free dynamic range (SFDR), defined at room temperature as follows [62]

$$\text{IMFDR (dB)} = \frac{2}{3}\left[\text{IIP3}_{Rec}\text{ (dBm)} + 174\,\text{dBm} - 10\log\left(\text{BW (Hz)}\right) - \text{NF}_{Rec}\text{ (dB)}\right], \quad (6.36)$$

where BW is the receiver bandwidth, defined by the application. The difference in IMFDR of the two receivers considered for the same application is thus independent of the bandwidth and is given by

$$\Delta\text{IMFDR (dB)} = \frac{2}{3}\left[\left(\text{IIP3}_{Rec,Psv} - \text{IIP3}_{Rec,Act}\right) - \left(\text{NF}_{Rec,Psv} - \text{NF}_{Rec,Act}\right)\right], \quad (6.37)$$

where $\text{IIP3}_{Rec,Act}$, $\text{IIP3}_{Rec,Psv}$, $\text{NF}_{Rec,Act}$ and $\text{NF}_{Rec,Psv}$ are the IIP3 points and noise figures of the *active* and *passive* receiver, respectively. Substituting the measured values for both receivers, it can observed that the *passive* receiver offers 6.2 dB wider dynamic range than the *active* one, even though it has 11.8 dB higher IIP3 point. The smaller improvement is due to the inherent factor $2/3$ in Eq. (6.37) and due to higher noise figure of the *passive* receiver.

6.3.2.4 Required LO Power

Measurement of conversion gain over LO power for an RF frequency of 24.01 GHz and an IF frequency of 10 MHz at an RF power of -20 dBm is given in Fig. 6.37.

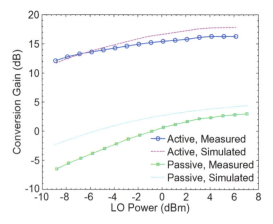

Fig. 6.37 Measured and simulated conversion gain over LO power.

6.3 Single-Channel Receivers

As can be observed, the *passive* receiver exhibits a higher dependence on the LO power than the *active* receiver. This is due to the stronger conversion loss dependence of the passive mixer on the LO power. The passive ring mixer requires sufficiently large power to switch the transistors on and off. Generating a minimal power of 6 dBm, required for sufficient performance of the *passive* receiver, might be a challenge, particularly if two passive ring mixers are integrated for in-phase and quadrature operation with a single VCO. Additionally, generating such high power levels leads to higher DC power consumption in the LO buffer stage, feeding the passive mixers. Taking into account a limited efficiency of the power amplifiers in CMOS in this frequency range, which is usually below 50%, the required DC power would exceed 8 mW. Thus, a *passive* receiver would lose the advantage of the lower DC power consumption compared to the *active* one, if VCO and power amplifier are also considered.

6.3.2.5 Isolation

The port-to-port isolation has been characterized for both circuits in measurement for input frequencies from 21.5 GHz to 26.5 GHz, as presented in Fig. 6.38.

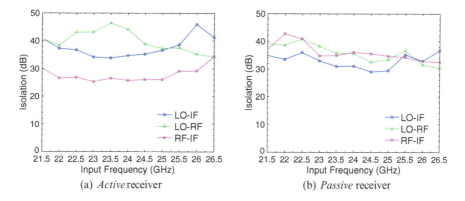

Fig. 6.38 Measured port isolation of the single-channel receivers.

All the isolation levels are sufficient, since the mixers used in both receivers are double-balanced and the reverse isolation offered by the LNA is high. The RF-IF isolation of the *active* receiver is much lower, because the RF signal is amplified in the Gilbert cell.

6.3.2.6 Temperature Performance

The circuit performance of both receivers has been characterized over temperature. The assembled test board has been placed into a Vötsch temperature chamber VT 4002. All the cables have been guided through a side wall opening. The RF and LO signals have been generated externally using two synthesized sweepers. The buffered differential IF output signals have been combined using a low-frequency hybrid ring coupler and observed on a spectrum analyzer. The temperature dependence of interconnect losses has been carefully accounted for using a power meter. The conversion gain as function of temperature, presented in Fig. 6.39, has been measured for an RF input frequency of 24.01 GHz, an IF frequency of 10 MHz, an RF power of −20 dBm and LO power of 3 dBm and 6 dBm for the *active* and the *passive* receiver, respectively.

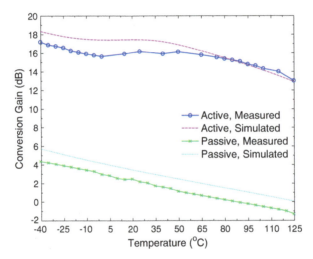

Fig. 6.39 Measured and simulated conversion gain as function of temperature.

Between −40 °C to 125 °C the gain drops by 4.2 dB and 5.7 dB for the *active* and *passive* receiver, respectively. It is interesting that between 0 − 50 °C the measured and simulated gain of the *active* receiver has a "plateau" region, whilst the gain of the *passive* receiver drops linearly with the temperature.

This behavior of the active mixer can be partially explained by a temperature self-compensation effect of the current mirror implemented here according to Fig. 6.13. Due to temperature dependent threshold voltage variation of −2.4 mV/°C [63], the reference branch current increases and the gate-source voltage of M7, M1 and M2 decreases with an increase in temperature. The transconductance g_m of transistors M1, M2 is proportional to the effective mobility μ_n and to the gate overdrive of M1 and M2. Therefore, when the gate overdrive of M1 and M2 increases as much as the mobility of M1 and M2 decreases with an increase in temperature [63], it is

6.3 Single-Channel Receivers

possible to compensate the transconductance, which is directly related to the voltage conversion gain of a Gilbert mixer, as shown in Eq. (6.21).

6.3.3 Receiver Results Summary and Comparison

The measurement results and comparison with state-of-the-art designs in CMOS using active and passive mixers are presented in Table 6.4. Both chips compare favorably to literature in terms of noise figure and chip area.

Table 6.4 Single-channel receiver performance comparison

Feature	[64]	[55]	Active	Passive	[31]	[65]
CMOS (μm)	0.18	0.18	0.13	0.13	0.13	0.13
f_T (GHz)	60	100	100	100	100	100
Integration	LNA,IFA Act Mix	LNA,IFA Act Mix	LNA,IFA Act Mix	LNA,IFA Psv Mix	LNA,IFA Psv Mix	LNA,Mix Psv Mix
f_{RF}/f_{IF} (GHz)	24/4.82	21.8/4.9	24/0.01	24/0.01	23.1/0.03	24/0.02
Gain (dB)	28.4	27.5	16	2	3.2	20.7
DSB NF(dB)	6	7.7	5	7.5	10	10.8
IP1dB (dBm)	-23.2	-23	-24.6	-13.4	-12.7	-23.3
IIP3 (dBm)	-13.0	NA	-15.8	-4	NA	-12.6
LO-IF (dB)	38.9	NA	34.6	31	NA	NA
RF-IF (dB)	20	NA	25.5	36	NA	NA
LO-RF (dB)	42.6	NA	44	35.5	62[1]	44
P_{dc} (mW)	54	30[2]	22.2	18.3	13.6	33.8[2]
V_{dc} (V)	1.8	1.5	1.5	1.5	1.6	1.2
Size (mm^2)	1.32	0.72	0.5	0.37	0.58	1.53

[1] Isolation is given for 2LO to RF, since a sub-harmonic mixer is used.
[2] Excluding IF amplifiers (IFA).

Comparing the two receivers, the *active* one offers higher gain, lower noise figure above the flicker noise corner frequency, lower LO power dependence, lower required LO power and exhibits better temperature stability. The *passive* receiver has the advantage of better input-referred linearity, wider intermodulation free dynamic range and also a much lower flicker noise. Both circuits have comparable chip area, current consumption and isolation.

The *passive* receiver is more suitable for direct down-conversion or for very low-IF radar realizations, requiring low flicker noise. However, in case of several receive channels integrated with a single transmitter in CMOS, the achievable LO power in CMOS might be insufficient to drive multiple passive mixers. Furthermore, the higher LO power would require higher DC power consumption of the LO buffers. Therefore, in such applications, the *active* receiver could be advantageous. Thus, further receiver designs in this work use active mixers.

6.4 IQ Receivers

The LNAs and active mixers in CMOS and SiGe technologies described in previous sections are now further integrated into receivers having In-phase and Quadrature (IQ) channels. The implementation of an IQ receive path provides unambiguous phase data, required for frequency-stepped continuous wave or for pulse radars. The down-converted signal can be treated as complex demodulated.

As mentioned previously, several publications report integrated single-channel receivers in CMOS consisting of a single LNA and a mixer [55], [64], whilst others report integration of complex large area phased arrays for wireless communications [66]. Also numerous works have been published on receivers in SiGe technology for applications around 24 GHz. Several publications focus e.g. on low noise figure [24] or on wide bandwidth [67]. Only a few publications report fully ESD-protected receivers [68].

This section presents IQ receivers in CMOS and SiGe, described in [69] and [70], respectively. Both circuits have been optimized for a low noise performance and lowest chip area. Furthermore, the SiGe receiver is designed for high ESD protection and temperature robustness, necessary for automotive applications. Section 6.4.1 describes the circuit design of the receivers. Section 6.4.2 presents the measurements of both circuits. Finally, section 6.4.3 summarizes the results and compares the presented receivers with the recently published works.

6.4.1 Design of IQ Receivers

6.4.1.1 IQ Receiver in CMOS Technology

The block diagram of the CMOS receiver is presented in Fig. 6.40. The circuit

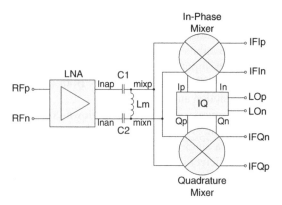

Fig. 6.40 CMOS IQ receiver block diagram.

6.4 IQ Receivers

includes an LNA, two active mixers and an on-chip In-phase and Quadrature (IQ) phase generation from an external Local Oscillator (LO) signal. The capacitors C1 and C2 along with inductor Lm compose a differential matching network between the LNA and the two mixers. Furthermore, the series capacitors of the matching network are also used for AC coupling between the stages.

The CMOS LNA has a similar topology to the one described in section 6.1.1. The input stage has been adjusted for improved input matching and stability under the modified load conditions by optimizing the input inductors L1 and L2 in Fig. 6.1. In addition to the ESD protection at the RF input, the internal virtual ground nodes between inductors L3 to L4 and L5 to L6 in Fig. 6.1 have been used for attaching capacitors in order to improve the common-mode stability of the receiver, as explained in section 6.1.1.

The CMOS active mixer is similar to the one described in section 6.2.1.1. The series inductors at the gates of M1 and M2 in Fig. 6.13 have been omitted, since sufficient inter-stage matching has been achieved using the aforementioned differential matching network. Thus, considerable chip area saving has become possible. The width of transistors M1, M2 has been correspondingly optimized to achieve similar noise matching as in the stand-alone circuit.

The schematic diagram of the polyphase filter for IQ generation is presented in Fig. 6.41. This solution has the advantage of considerable area saving and simplicity compared to such alternatives as an oscillator at the double frequency or on-chip hybrid quadrature couplers. The drawback of introducing higher losses when loaded by finite impedance is less crucial for sufficient LO power. The topology for better phase balance over wider frequency range [71] has been chosen and dimensioned for a center frequency f_0 of 24 GHz using the simple relation $f_0 = 1/(2\pi RC)$.

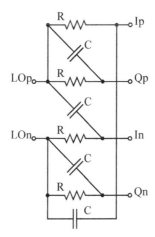

Fig. 6.41 Schematic diagram of the polyphase filter.

6.4.1.2 IQ Receiver in SiGe Technology

The block diagram of the SiGe receiver is presented in Fig. 6.42.

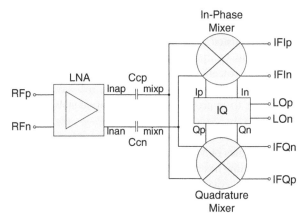

Fig. 6.42 SiGe IQ receiver block diagram.

The SiGe receiver also includes an LNA, two active mixers and a polyphase network that generates quadrature phase of an input LO signal for the mixers. The capacitors Ccp and Ccn are used for AC coupling between the LNA and the following mixer stages.

The LNA and mixer building blocks are very similar to the ones described in section 6.1.2 and 6.2.1.2, respectively. The output of the LNA has been designed to match 100 Ω differentially. Also the differential mixer input has been designed to match 100 Ω. However, since the inputs of two mixers are now seen in parallel at the output of the LNA, different impedance transformation is required. Therefore, the interstage matching has been optimized by modifying the impedance transformation ratio of the tapped capacitor LC tanks at the output of the LNA.

Similarly as in case of the CMOS receiver, the polyphase network presented in Fig. 6.41, has been implemented for IQ generation in this case. The filter components have been dimensioned very similarly.

Sufficient ESD robustness and performance stability over a wide range of temperatures are required for hostile environment such as in automotive applications. Therefore, ESD protection diodes have been attached directly at each node, similarly as presented in Fig. 6.3, of the RF input, the IF output and the LO input ports. The partial temperature compensation has been realized in the current mirror of the LNA and mixer blocks, as described in sections 6.1.2 and 6.2.1.2, respectively.

In both CMOS and SiGe receivers blocking capacitors and ESD power clamps have been added between the power supply and ground.

6.4.2 IQ Receiver Measurements

The CMOS and SiGe receiver circuits have been processed in Infineon's C11N and B7HF200 technologies, respectively. The annotated chip micrographs of the IQ receivers are presented in Fig. 6.43. The size including the pads is 0.48 mm^2 for CMOS and 1 mm^2 for the SiGe chip, respectively. As can be observed in Fig. 6.43(a), the capacitors Cs1, Cs2 are attached at the symmetry axis in order to improve the common-mode stability.

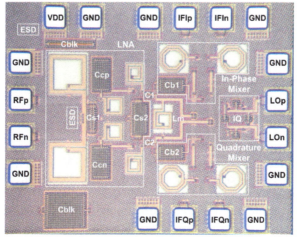

(a) CMOS (0.78 mm × 0.61 mm)

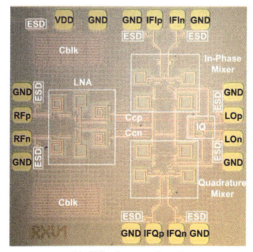

(b) SiGe (1 mm × 1 mm)

Fig. 6.43 Chip micrographs of the IQ receiver circuits.

As described in previous sections, the CMOS chip was thinned to 185 µm and the SiGe chip to 350 µm. For characterization, the chips have been mounted on a test PCB shown in Fig. 6.44.

Fig. 6.44 Differential IQ receiver characterization board.

The CMOS receiver consumes only 16 mA from 1.5 V, whilst the SiGe receiver consumes 39 mA from a single 3.3 V supply. Both chips have been characterized for RF input frequencies from 21.5 GHz to 26.5 GHz, whilst the IF was kept constant at 10 MHz. The input LO power was 3 dBm for the CMOS and 0 dBm for the SiGe chip. The input RF power was −40 dBm. The measured and simulated conversion gain and noise figure versus RF frequency are presented in Fig. 6.45 for CMOS and in Fig. 6.46 for SiGe, respectively.

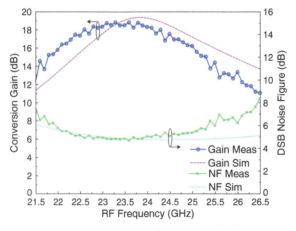

Fig. 6.45 Measured and simulated gain and noise figure of CMOS IQ receiver.

6.4 IQ Receivers

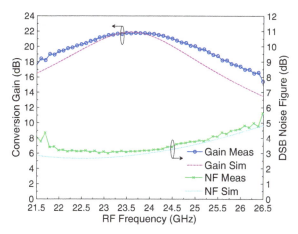

Fig. 6.46 Measured and simulated gain and noise figure of SiGe IQ receiver.

As can be observed, both circuits achieve a very low DSB noise figure of 5 dB for CMOS and 3.1 dB for SiGe around the center frequency of 24 GHz. The bipolar circuit offers a higher gain of 21.5 dB compared to 18 dB of CMOS.

The linearity measurements were performed for an LO power of 3 dBm for the CMOS and 0 dBm for the SiGe chip, respectively. The IP1dB was measured to be −26.8 dBm for CMOS and −20.5 dBm for SiGe at an RF frequency of 24.01 GHz and an IF frequency of 10 MHz. The IIP3 measurement has been performed by applying 24 GHz at the LO port and two input tones 24.01 GHz and 24.011 GHz at the RF port. The CMOS and SiGe receivers exhibit IIP3 of −17.7 dBm and −11 dBm, respectively. The IP1dB and IIP3 linearity measurements of the receivers are presented Fig. 6.47 and Fig. 6.48.

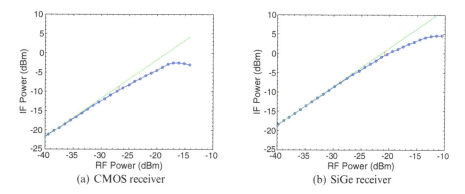

(a) CMOS receiver

(b) SiGe receiver

Fig. 6.47 IP1dB measurement of the IQ receivers in CMOS and SiGe.

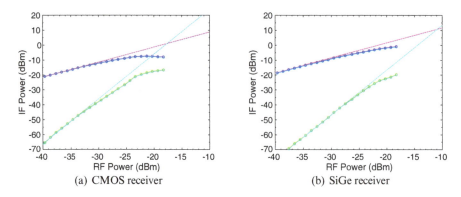

Fig. 6.48 IIP3 measurement of the IQ receivers in CMOS and SiGe.

The conversion gain of the CMOS and SiGe IQ receivers has been measured as function of LO power at an RF frequency of 24.01 GHz, an IF frequency of 10 MHz and an RF power of −40 dBm, as presented in Fig. 6.49.

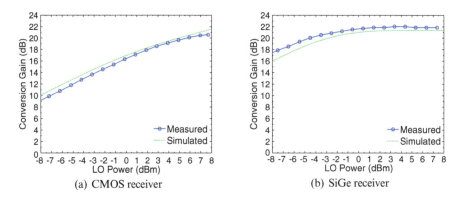

Fig. 6.49 Conversion gain of the IQ receivers over LO power.

It can be observed that LO power levels of 3 dBm and 0 dBm are sufficient to achieve optimal performance in CMOS and SiGe, respectively. Further can be observed that the CMOS receiver exhibits a higher dependence on the LO power than the SiGe receiver. This is due to the stronger dependence on LO power of the CMOS mixer compared to the SiGe mixer. It shall be considered that the differential pairs comprising the LO switching stage M3-M4, M5-M6 in Fig. 6.13 for the CMOS mixer and N3-N4, N5-N6 in Fig. 6.14 for the SiGe mixer are required to commutate the current from the RF transconductance stage rapidly either to the positive or negative output depending on the differential LO input signal. In this design, the LO stage pairs have been biased such that the differential transconductance is 4 mS and

6.4 IQ Receivers

44 mS in the CMOS and in the SiGe mixer, respectively. The slope of the differential large-signal transfer characteristics of a differential pair in the SiGe mixer is much steeper compared to the CMOS mixer. Thus, a smaller differential input voltage swing at the LO input is sufficient to fully commutate the current in the SiGe active mixers compared to the CMOS active mixers [72].

The isolation between the ports LO-RF, LO-IF and RF-IF has been characterized for both receivers over input frequencies 21.5 GHz to 26.5 GHz, as shown in Fig. 6.50. As can be seen, a very good isolation has been achieved for both receivers. The isolation levels between the ports LO-RF, LO-IF and RF-IF at the center frequency of 24 GHz were measured to be 55 dB, 37 dB and 28 dB, for the CMOS receiver, and 51 dB, 38 dB and 19 dB for the SiGe receiver, respectively. The RF-IF isolation is lower in both receivers since the input signal is amplified in the LNA and in the active mixer.

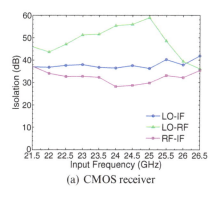
(a) CMOS receiver

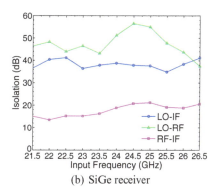
(b) SiGe receiver

Fig. 6.50 Measured port isolation of the receivers.

The performance stability of the receivers has been verified in measurement over a wide range of temperatures from −40 °C to 125 °C, as shown in Fig. 6.51. The gain drops over the temperature range from the room temperature of 25 °C to the temperature of 125 °C by 6 dB and 3.5 dB for CMOS and SiGe receivers, respectively. The noise figure increases over this temperature range by 3.5 dB and 2.3 dB for the CMOS and SiGe receivers, respectively. The SiGe receiver exhibits much better temperature stability. The temperature stability of the CMOS IQ receiver is degraded compared to the CMOS single-channel receiver presented in section 6.3.2.6 due to several reasons. Firstly, the IQ receiver is more sensitive to temperature variations than the single-channel receiver, since in the design of the mixer the series input inductors have been omitted, as described in section 6.4.1.1. The interstage matching of the IQ receiver has been achieved using a differential matching network and by appropriate choice of the width and biasing of the RF transconductance stage transistors. However, the biasing conditions and thus the input impedance at the RF port of the mixer change with temperature, thus the interstage matching gets worse and causes stronger performance variation with temper-

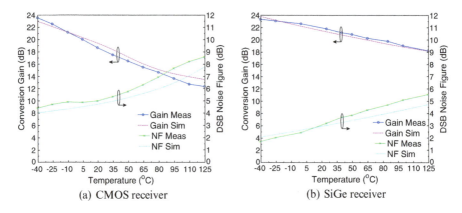

Fig. 6.51 Measured gain and noise figure of IQ receivers versus temperature.

ature in this circuit. Therefore, omitting the series inductors has reduced area at the expense of temperature stability. Secondly, even though both receivers have been measured using 3 dBm LO input power, the actual power at the LO ports of the mixers is lower in the IQ receiver due to the insertion loss of the polyphase filters. At higher temperatures the differential transconductance of the LO switching stage decreases and a higher LO drive voltage is required. Thus, the lower actual LO drive voltage becomes more critical in the CMOS IQ receiver.

The I/V and spot leakage curves of the ESD TLP stress between RF input and GND of the SiGe IQ receiver with 100 ns wide pulses having rise time of 10 ns is presented in Fig. 6.52 for a negative pulse. The minimum failure current on the high-frequency pins was observed to be above 1.5 A. This corresponds to HBM ESD hardness higher than 2 kV according to JESD22-A114 standard.

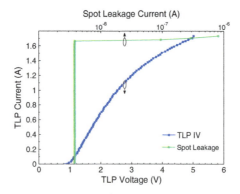

Fig. 6.52 TLP measurement of the RF input to GND for the SiGe IQ receiver.

6.4.3 IQ Receiver Results Summary and Comparison

The measurement results and comparison with the recently published 24 GHz receivers in CMOS and SiGe are presented in Table 6.5. According to author's knowledge, both presented receivers offer the lowest reported noise figure to date. Furthermore, the CMOS receiver offers the lowest chip area, better isolation and a low current consumption. Additionally, the SiGe receiver compares favorably to literature in terms of linearity and chip area. The SiGe receiver also offers ESD protection above 2 kV on high-frequency pins.

Table 6.5 IQ receiver performance comparison

Feature	[64]	[55]	CMOS	SiGe	[24]	[67]
Technology	0.18 μm CMOS Si	0.18 μm CMOS Si	0.13 μm CMOS Si	0.35 μm Bipolar SiGe	0.17 μm BiCMOS SiGe	0.18 μm BiCMOS SiGe
f_T (GHz)	60	100	100	200	170	130
Integration	LNA,Mix IFA	LNA,Mix IFA	LNA,IQ 2xMix	LNA,IQ 2xMix	LNA,Mix IFA	2xLNA,2xMix VCO, IFA
Topology	SE	SE	Diff	Diff	Diff	Diff
f_{RF}/f_{IF} (GHz)	24/4.82	21.8/4.9	24/0.01	24/0.01	24/0.1	24/0.02
Gain (dB)	28.4	27.5	18	21.5	31.8	39
DSB NF(dB)	6	7.7	5	3.1	3.5	7
IP1dB (dBm)	-23.2	-23	-26.8	-20.5	-40	NA
IIP3 (dBm)	-13.0	NA	-17.7	-11	NA	-34
LO-IF (dB)	38.9	NA	37	38	33	NA
RF-IF (dB)	20	NA	28	19	NA	NA
LO-RF (dB)	42.6	NA	55	51	75	NA
P_{dc} (mW)	54	30^2	24	129	80^2	920^1
V_{dc} (V)	1.8	1.5	1.5	3.3	1.2/2.5	4
Size (mm^2)	1.32	0.72	0.48	1	0.58	2.55^1

[1] Additional components are included.
[2] Excluding buffers.

Very low noise figures have been achieved for both receivers, since the overall NF is dominated only by the NF of the LNA due to the very low mixer noise figure.

In comparison to the CMOS receiver, the SiGe receiver provides higher gain, lower noise figure and better linearity at the expense of higher power consumption. Furthermore, the SiGe circuit offers much better performance stability than CMOS over a wide temperature range. To author's knowledge this is the first work comparing two 24 GHz receivers designed in CMOS and SiGe technologies.

The presented CMOS receiver is more suitable for low-cost 24GHz radar applications, whilst the presented SiGe receiver is more suitable for integration in radar sensors for robust automotive or industrial applications in the 24 GHz ISM band.

6.5 Integrated Passive Circuits

As mentioned previously, the differential signaling gains popularity for high-frequency circuit design. However, since many applications require single-ended inputs and since measurement equipment usually has single-ended ports, a single-ended to differential signal conversion has to be performed in the system. This can be realized e.g. using a 180° hybrid ring coupler with a terminated isolation port or using a lattice-type LC-balun that should ideally provide an equal power split and 180° phase difference across the differential port. In the previous sections this conversion has been realized on-board using hybrid ring couplers, as e.g. shown in Fig. 6.5. However, a compact on-chip realization of an equal power split with 180° phase shift may be very advantageous to reduce the module size and to increase the integration level of the system on chip.

Additionally, most microwave systems require on-chip quadrature generation for efficient modulation techniques in communication applications or for unambiguous phase data evaluation in radar applications. This requires components that offer equal power split and 90° phase difference. One on-chip realization option is using a 90° hybrid quadrature coupler with a terminated isolation port.

Numerous publications report integrated baluns or quadrature couplers. A highly popular concept is to use a Wilkinson divider along with phase shifters formed by transmission lines that provide the necessary phase difference at the output [73]. However, this concept is more suitable for millimeter-wave frequencies above 60 GHz, whilst at 24 GHz the components would consume a very large chip area. Thus, at this frequency it is more suitable to realize filters using lumped on-chip elements, as spiral inductors and capacitors.

Therefore, this section deals with the design of narrow-band 24 GHz passive on-chip lumped-element hybrid components that provide phase difference of 90° and 180°. Section 6.5.1 presents the design of a compact on-chip 180° power splitter/combiner based on a lattice-type LC-balun, of a hybrid power splitter/combiner based on a lumped-element 90° quadrature coupler and of a lumped-element hybrid 180° ring coupler. Section 6.5.2 presents the measurement results of the aforementioned components. Finally, section 6.5.3 summarizes and discusses the presented results.

6.5.1 Circuit Design and Layout Considerations

6.5.1.1 On-Chip 180° Power Splitter/Combiner

The conceptual schematic diagram of a 180° power splitter/combiner based on a lumped lattice-type LC-balun [74] is presented in Fig. 6.53(a). This circuit transforms the single-ended input to differential outputs by shifting the phase in one branch by +90° and shifting the phase in another branch by -90°. At the angular resonance frequency of

6.5 Integrated Passive Circuits

$$\omega_0 = \frac{1}{\sqrt{LC}}, \tag{6.38}$$

where L and C are the inductance and capacitance of the balun components, the unbalanced input resistance R_{in} is transformed into the balanced output resistance R_{out} given by

$$R_{out} = \frac{L}{C} \cdot \frac{1}{R_{in}}. \tag{6.39}$$

The classical LC-balun in Fig. 6.53(a) can be thus dimensioned by combining Eq. (6.38) and (6.39) as follows

$$L = \frac{\sqrt{R_{in} \cdot R_{out}}}{\omega_0},$$

$$C = \frac{1}{\omega_0 \sqrt{R_{in} \cdot R_{out}}}. \tag{6.40}$$

The inductors can be combined into a transformer, as in Fig. 6.53(b), in order to allow a reduced-size physical implementation.

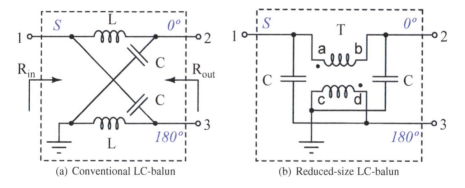

Fig. 6.53 Schematic diagram of lattice-type LC-balun implementations.

The presented balun has been designed for a single-ended impedance of 50 Ω and a differential impedance of 100 Ω. Typically, lower values of capacitors are required than expected in theory, due to the parasitic capacitance of the on-chip coils to the substrate. Thus, an accurate modeling of components has to be performed in simulation and the component values need to be modified accordingly. The values obtained from (6.40) are used as initial values for the design.

The required series inductance of the first and second coil of the transformer T in Fig. 6.53(b) is multiplied by approximately $1/(1+k)$, where k is the coupling coefficient. Realizing T as an interleaved transformer allows to achieve a moderately high coupling factor of about 0.7 [17]. A possible layout implementation, realized in this design, is presented in Fig. 6.54. The inherent asymmetry of the structure results in a difference of about 20 pH between the primary and secondary winding

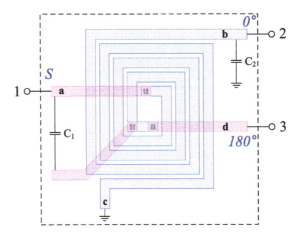

Fig. 6.54 Implementation of an inductively-coupled reduced-size LC-balun.

inductances of this implementation, as has been observed in simulation. Thus, the capacitors C_1 and C_2 have different values. Diagonal routing of the branch allows to attach the capacitor C_1 between the ports 1 and 3.

6.5.1.2 On-Chip 90° Power Splitter/Combiner

The schematic diagram of a lumped-element 90° hybrid branch-line coupler based splitter/combiner is presented in Fig. 6.55(a). In order to minimize the chip area, the inductors can be combined into a transformer, as described in Fig. 6.55(b).

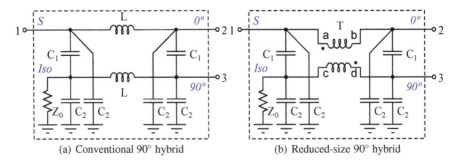

(a) Conventional 90° hybrid (b) Reduced-size 90° hybrid

Fig. 6.55 Schematic diagram of lumped-elements 90° hybrid implementations.

This circuit is based on a lumped-element realization of the classical branch-line coupler. The distributed realization requires $\lambda/4$ transmission lines, which are replaced by lumped-element equivalents based on simple π-type filters [53]. At reso-

6.5 Integrated Passive Circuits

nance frequency these LC filters exhibit equivalent characteristics to a transmission line with characteristic impedance Z_0 and phase delay ϕ. The combination of low-pass and high-pass LC filters provides 90° phase difference between the outputs 2 and 3 and signal cancellation at the isolation node. The circuit values in Fig. 6.55(a) can be dimensioned using [53]

$$L = \frac{Z_0}{\omega_0 \cdot \sqrt{2}},$$
$$C_1 = \frac{1}{\omega_0 \cdot Z_0}, \qquad (6.41)$$
$$C_2 = \frac{1}{\omega_0^2 \cdot L} - C_1,$$

where Z_0 is the port impedance, which is 50 Ω in this design, and $\omega_0 = 2\pi f_0$ is the angular operation frequency, with f_0 of 24 GHz here. Again, the theoretical model gives only the initial values for the design. The required values of C_2 in Fig. 6.55(b) are smaller than estimated due to capacitance of the coils to the substrate. For a highly conductive substrate the four C_2 capacitors may be omitted.

The coupling factor of about 0.7, achieved using the interleaved transformer, is very close to the desired value of 0.707, required theoretically for an equal power split of the coupler [75]. The realized layout implementation is presented in Fig. 6.56. In this case the capacitance value of C_1 was very small and it has been physically attached between the coupled inductor nodes.

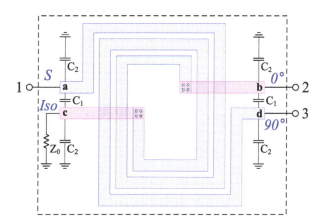

Fig. 6.56 Layout implementation of the inductively-coupled 90° power splitter.

6.5.1.3 On-Chip 180° Hybrid Ring Coupler

The schematic diagram of a 180° hybrid ring coupler, also commonly referred to as the *rat-race* coupler, realized using lumped elements is shown in Fig. 6.57.

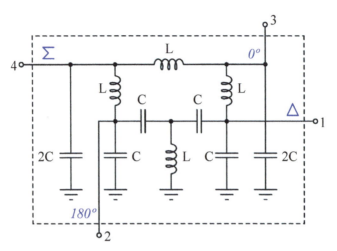

Fig. 6.57 Schematic diagram of a 180° hybrid ring coupler implementation.

This circuit is the lumped-element equivalent of the classical ring coupler. Similarly to the previous section, the $\lambda/4$ lines have been replaced by low-pass π-type filters, whilst the long $3/4\lambda$ section has been replaced by a high-pass π-type filter. All lines on the ring have the characteristic impedance of $\sqrt{2}Z_0$, where Z_0 is the port impedance, which is again 50 Ω here. The circuit is dimensioned at $\omega_0 = 2\pi f_0$, where f_0 is the center frequency of 24 GHz using Eq. (6.27).

This circuit implementation has the interesting feature that power fed at port 1 is split nearly equally between ports 2 and 3, whilst the phase difference between the ports 2 and 3 is close to 180° over a wider frequency range rather than just at a single frequency. This is due to flat frequency response of the transmission S-parameters around the center frequency inherent to this realization, as can be also observed in [76] that has originally proposed this circuit topology.

Unlike the LC-balun described in section 6.5.1.1, this circuit also provides an isolated port, in addition to equal power split and 180° phase difference between outputs. This is particularly useful for several applications in order to isolate signals. For example, in the mixer presented in section 6.2.2.2, the LO and RF signals are applied to the mutually isolated ports of the ring coupler.

The lumped-element 180° hybrid ring coupler requires four on-chip inductors, thus consumes larger area than the LC-balun. The layout implementation in this design follows closely the schematic diagram in Fig. 6.57, thus the component arrangement is annotated directly on the chip photo in the next section.

6.5 Integrated Passive Circuits

6.5.2 Realization and Measurement Results

The presented lumped-element approaches are applicable to any semiconductor technology that offers integrated inductors and capacitors with sufficiently high quality factors, typically above 10, in the interesting frequency range. As a proof of concept, the passive circuits have been realized in Infineon's B7HF200 process.

The structures have been characterized on-wafer using Cascade Microtech Infinity probes up to 40 GHz with 100 μm pitch in GSSG configuration and a four-port VNA. Accurate four-port SOLT calibration has been performed to set the reference planes at the pads of the test chips. Additionally, a two-step Open-Short de-embedding technique has been implemented in order to move the reference planes from the pads to the terminals of the structure.

6.5.2.1 On-Chip 180° Power Splitter/Combiner

The annotated chip micrograph of the 180° power splitter/combiner based on an LC-balun realized using lumped-elements is presented in Fig. 6.58. The chip size of the circuit excluding pads is 0.018 mm².

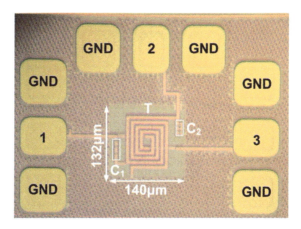

Fig. 6.58 Chip micrograph of 180° power splitter (size 140 μm × 132 μm).

The measured and simulated S-parameters of the LC-balun based 180° power splitter/combiner are presented in Fig. 6.59. Due to high sensitivity to parasitics, the measured equal power split frequency has been observed at 22.5 GHz. The insertion loss added to the 3 dB at the point of equal power split is 1.9 dB.

The phase difference between the ports, defined as $\Delta\phi = |\angle S_{31} - \angle S_{21}|$, and the amplitude imbalance, defined as $\Delta S (\text{dB}) = S_{31} (\text{dB}) - S_{21} (\text{dB})$, of the 180° power splitter/combiner is depicted in Fig. 6.60. Amplitude and phase imbalance of 0.6 dB and 10° have been observed at 24 GHz.

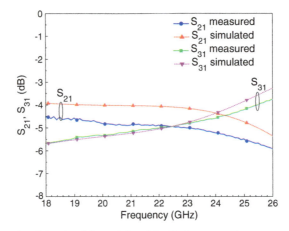

Fig. 6.59 Measured and simulated S_{21} and S_{31} of the 180° power splitter.

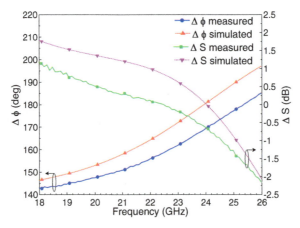

Fig. 6.60 Measured and simulated imbalance of the 180° power splitter.

The measured and simulated reflection S-parameters of the single-ended port 1 and the differential port comprising the ports 2 and 3 are presented in Fig. 6.61. The differential reflection S-parameter, defined in the given port nomenclature as $s_{dd} = \frac{1}{2}(s_{22} - s_{23} - s_{32} + s_{33})$, is considered here to verify differential matching to 100 Ω. As can be observed, good matching has been achieved at both ports.

6.5.2.2 On-Chip 90° Power Splitter/Combiner

The annotated chip micrograph of the branch-line coupler based 90° power splitter/combiner is presented in Fig. 6.62. The chip size of the circuit excluding pads is 0.018 mm^2.

6.5 Integrated Passive Circuits

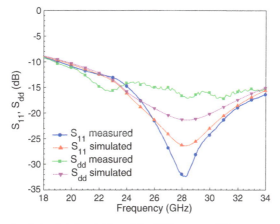

Fig. 6.61 Measured and simulated matching of the 180° power splitter.

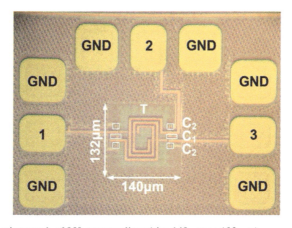

Fig. 6.62 Chip micrograph of 90° power splitter (size 140 μm × 132 μm).

The measured and simulated S-parameters of the 90° power splitter/combiner are presented in Fig. 6.63. The frequency of the equal power split has been observed at 26.5 GHz. The added insertion loss of the device at this frequency is 0.5 dB.

The phase difference between the ports and the amplitude imbalance of the 90° coupler based power splitter/combiner are presented in Fig. 6.64. Amplitude and phase imbalance of 2 dB and 0.6° have been observed at 24 GHz.

The measured reflection S-parameters of the structure are presented in Fig. 6.65. As can be observed, good port matching of about −20 dB has been achieved at the center frequency of 24 GHz on all ports.

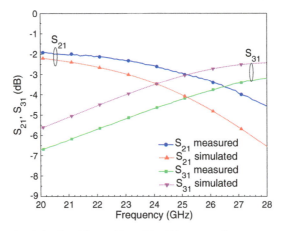

Fig. 6.63 Measured and simulated S_{21} and S_{31} of the 90° power splitter.

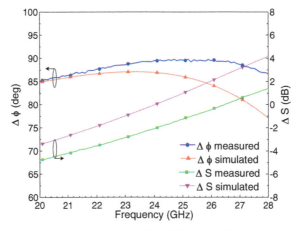

Fig. 6.64 Measured and simulated imbalance of the 90° power splitter.

6.5.2.3 On-Chip 180° Hybrid Ring Coupler

The annotated chip micrograph of the lumped-element 180° hybrid ring coupler is presented in Fig. 6.66. The chip size of the circuit excluding pads is 0.044 mm^2.

The measured and simulated transmission S-parameters of the 180° hybrid ring coupler are presented in Fig. 6.67. The circuit has been designed in simulation for an equal power split at 24 GHz. However, as explained previously, this topology provides wider overlap of the transmission S-parameters around the center frequency and thus there is an additional intersection around 31 GHz. In measurement the lower intersection frequency is shifted to 22 GHz and the upper to 32 GHz. The added insertion loss at the points of equal power split is 1.5 dB.

6.5 Integrated Passive Circuits 141

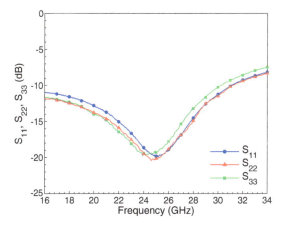

Fig. 6.65 Measured reflection S-parameters of the 90° power splitter.

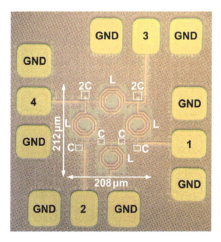

Fig. 6.66 Chip micrograph of 180° hybrid ring coupler (size 208 μm × 212 μm).

The phase and amplitude differences between the ports are depicted in Fig. 6.68.

The deviation from 180° does not exceed ±5°, whilst the amplitude imbalance remains below 3.4 dB over a very wide frequency range from 23 GHz to 40 GHz.

The port isolation between the ports 1 and 4, also referred to as the difference Δ and summation Σ ports, respectively, is shown in Fig. 6.69. As can be observed, good isolation has been achieved over the entire frequency range. As mentioned previously, this is particularly important for mixer applications.

The measured reflection S-parameters of the 180° hybrid ring coupler are presented in Fig. 6.70. Good port matching of better than −15 dB has been achieved

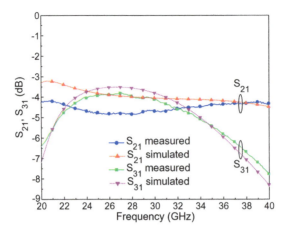

Fig. 6.67 Measured and simulated S_{21} and S_{31} of the 180° hybrid coupler.

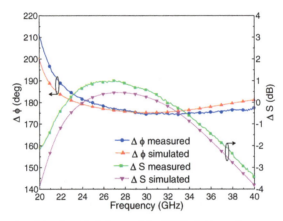

Fig. 6.68 Measured and simulated imbalance of the 180° hybrid coupler.

close to the center frequency of 24 GHz on all ports. However, the port matching degrades significantly at higher frequencies.

6.5 Integrated Passive Circuits

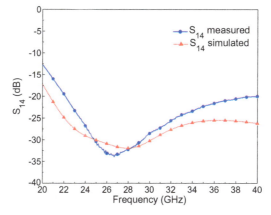

Fig. 6.69 Measured and simulated port isolation of the 180° hybrid coupler.

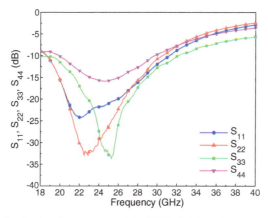

Fig. 6.70 Measured reflection S-parameters of the 180° hybrid coupler.

6.5.3 Results Summary and Discussion

This section has presented three passive integrated components that can be used for power splitting/combining with 180° or 90° phase difference between the output ports. The key parameters of these circuits are summarized in Table 6.6. The 90° power splitter offers a very low added insertion loss and excellent phase balance, but moderate amplitude imbalance. The 180° components offer a very good amplitude balance, but higher added insertion loss. All the circuits offer good port matching at 24 GHz and consume small chip area.

It has been observed that the described lumped-element realizations are highly sensitive to parasitics and component value tolerances. Thus, particular attention is

Table 6.6 Performance of on-chip passive components around 24 GHz

Feature	180° LC-Balun	90° Splitter	180° Ring
Added Insertion Loss (dB)	1.9[1]	0.5[2]	1.5[3]
Amplitude Imbalance (dB)	0.6	2	0.6
Phase Imbalance (°)	10	0.6	2
Port Matching (dB)	-14	-18	-16
Port Isolation (dB)	NA	NA	-26
Size (mm^2)	0.018[4]	0.018[4]	0.044[4]

[1] At the equal power split frequency of 22.5 GHz.
[2] At the equal power split frequency of 26.5 GHz.
[3] At the equal power split frequency of 22 GHz.
[4] Excluding pads.

required to accurate modeling of the components and interconnect parasitics during the design. Minor changes of component values might shift the resonance frequency by several gigahertz, since the resonance frequencies of the low-pass and high-pass filters, used in all these realizations, might diverge.

The LC-balun based 180° power splitter/combiner is suitable for integration at the RF and LO ports of the receivers presented in this work to provide single-ended to differential conversion. The performance degradation is not expected to be higher than 2 dB in gain and noise figure.

The 90° coupler based power splitter/combiner can be used for IQ generation in the receivers. However, the polyphase filters consume a much lower area and offer excellent amplitude and phase balance, at the expense of increased losses.

The 180° hybrid ring coupler offers good isolation, but requires larger area than the LC-balun. Thus, the ring coupler is more suitable for applications that require port isolation in addition to the equal 180° power split.

The presented passive components can be implemented for further development and on-chip integration of the 24 GHz radar front-end circuits.

6.6 Circuit-Level RF ESD Protection

ESD (electrostatic discharge) events, that are potentially destructive for electronic devices, may occur in integrated circuits under numerous circumstances. This can happen for example in the manufacturing stage during handling by pick and place, electrical testing or packaging or also very often during human handling of the chips. During contact between a charged body (typically charged to a level of several kilovolts or higher) with an integrated circuit (IC) high voltage or current discharge takes place and causes temporarily large electric fields and high current densities in the device. These transient ESD pulse events may result in increased leakage or breakdown of insulators and/or thermal damage of ICs. The ESD subject becomes particularly important in sub-micrometer silicon technologies due to low oxide breakdown voltages of devices.

In order to be able to determine the robustness of the ICs to ESD events, various testing standards have been developed to emulate different types of events that may occur: human body model (HBM), machine model (MM), charged device model (CDM) or socketed device model (SDM). Besides the described models, the transmission line pulse (TLP) technique, that provides not only the failure level, but also the dynamic behavior of protection devices, has become very popular.

ESD protection strategies can be divided into two categories: external and internal protection [77]. External protection includes for example proper grounding and shielding, maintaining static-free working environment and appropriate anti-static storage of components. This is meant to prevent static charge generation during wafer fabrication, human handling, packaging or system assembly. Internal protection is realized using on-chip protection networks that provide a robust low-impedance discharge path for large ESD currents to flow. Thus, the high current ESD transients are "diverted" from the sensitive internal circuitry. Internal protection can be realized in numerous ways. There has been a great deal of research on the protection devices and circuit techniques for providing ESD protection of circuits. A very good summary of the state-of-the-art ESD protection devices, circuit-level techniques and measurement methods is given in the following book [78].

During normal operation of the IC, the ESD protection networks should have a very high impedance and not affect the functionality of the protected circuitry. This is a crucial requirement, which is particularly hard to fulfill for efficient ESD protection of microwave circuits, since the most commonly used ESD protection approach is to attach ESD devices directly at the circuit pads. However, ESD protection devices usually introduce a significant capacitance and thus deteriorate the high-frequency performance of the protected circuit. Numerous circuit-level techniques have been developed that either minimize, take advantage of or avoid this parasitic capacitance. A brief overview of the existing circuit-level ESD protection techniques for RF circuits is given in section 6.6.1. Sections 6.6.2 and 6.6.3 describe further concepts proposed and analyzed in this work.

6.6.1 Overview of Circuit-Level Protection Techniques

The global ESD protection scheme of an integrated circuit has to be carefully planned to provide a safe discharge path for all possibilities of ESD events between any two pads on the chip. As mentioned previously, ESD protection networks have to provide a low-impedance path from input pads to ground in order to "divert" a large current during an ESD event. The networks are also required to prevent over-voltage at the transistors and thus to clamp the voltage level at the pads. Thus, ESD protection networks commonly comprise semiconductor devices such as e.g. Zener diodes, *pn* diodes, MOS or bipolar transistors and/or silicon-controlled rectifiers (SCRs) as ESD devices. Voltage-current characteristics of these devices exhibit a useful feature in common. Starting from a certain turn-on voltage, the voltage across the device is ideally independent of the current through it. In practice, a change in current causes a minimal change in voltage, resulting in a low non-zero incremental resistance. Therefore, ESD devices have the ability to conduct high current, whilst the voltage across the device remains low and nearly constant. The voltage is thus clamped during an ESD event at a sufficiently low level below the breakdown voltage of the protected transistors to allow safe operation of the internal circuit.

There are two main approaches during the design of a global ESD protection scheme: rail-based and local clamping [79]. In the rail-based approach typically *pn* diodes are attached at the input pads. The diodes are connected in reverse bias and thus do not conduct under normal circuit operation conditions. In case of an ESD event at the input pad, one of the diodes becomes forward biased and "forwards" the large transient current either to the power supply, where it is shunted by a power clamp device, or shorts it directly to the ground rail as shown in Fig. 6.71.

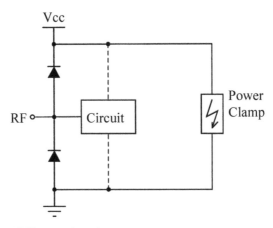

Fig. 6.71 Rail-based ESD protection scheme.

In case of a local clamping, the clamping device itself is attached directly at the pads and not between the rails. Thus, when an ESD stress pulse is applied between the RF pin and ground, the ESD current flows directly from the pad to ground, avoiding the excess voltage build-up on the rails.

In order to be able to withstand high currents and provide low ohmic resistance in the on-state, the ESD devices have to be physically large. However, this results in large parasitic capacitances. Using any of the two aforementioned global protection strategies, the highly capacitive devices have to be attached directly at the pads. Obviously, this is a major handicap for high-frequency circuits, since the capacitance between the pad and DC supplies poses a very low-impedance path for RF signals. Therefore, in the global ESD protection scheme of an integrated circuit the RF pins have to be treated separately.

6.6 Circuit-Level RF ESD Protection

There are several possibilities to approach this subject. The most common are called low-C ESD protection and ESD co-design [80]. The first refers to strong optimization of ESD structures for lowest possible capacitances. However, in order to reach acceptable capacitance values below 100 fF the ESD protection levels have to be compromised [79]. Thus, there is a trade-off between the size of ESD protection devices, protection level and parasitic effects.

The co-design approach is more sophisticated, since it takes the parasitics of the ESD device into account during the design stage of the RF circuit. The capacitive parasitics can be e.g. implemented as part of the impedance matching network [81], as part of an LC tank [82], in order to realize lumped capacitances as part of a transmission line [83] or even to resonate out a bondwire [84]. Furthermore, the parasitics of an ESD device can become transparent to the performance of the RF circuit by attaching them at the middle tap of a T-coil, whilst at the same time good broadband matching can be achieved [85]. Another idea of isolating the ESD parasitics from the RF signals by attaching the ESD devices at the virtual ground node of differential devices is also analyzed in this work, presented in [86] and discussed in section 6.6.2.

An additional approach proposes using inductive protection. This allows achieving very high protection levels at high frequencies, without using rectifying semiconductor based ESD protection devices apart from the power clamp [87]. Further development of this idea is to use a transformer at the input [88]. This idea has been further developed in this work and the transformer has been also used for matching and as a balun to convert between single-ended to differential signals, as discussed in section 6.6.3. This protection technique seems to provide the highest HBM protection levels on RF pads for circuits at microwave and millimeter-wave frequencies.

Most publications report co-design techniques demonstrated on single-ended circuits below 10 GHz. This work focuses on efficient protection of differential circuits. The techniques presented in sections 6.6.2 and 6.6.3 have been verified for differential LNAs at 18 GHz and 24 GHz.

6.6.2 Virtual Ground Concept

As mentioned throughout this work, most high-frequency integrated circuits use differential signaling. Every signal can be decomposed into a linear combination of common and differential modes. The symmetry axis introduces a constant potential for differential mode variations, thus it can be considered as a virtual AC ground, as illustrated in Fig. 6.72.

This means, large highly capacitive ESD protection structures, added at the symmetry line of the structure, would be transparent to the high-frequency signals and not impact the circuit performance. The same concept can be used for attaching further devices such as bypass capacitors for improving the common-mode circuit stability, as explained in section 6.1.1. The passive networks should be symmetrical and could be realized for example as inductors or transmission lines. The series re-

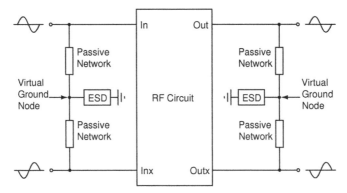

Fig. 6.72 Virtual ground ESD protection conceptual diagram.

sistance of these networks has to be very low to avoid large voltage drop and power dissipation.

The principle has been applied to a differential 18 GHz LNA in 0.13 μm CMOS, presented in [89]. The topology of this LNA is very similar to the one shown in Fig. 6.1 for the 24 GHz LNA described in detail in section 6.1.1. At the input and output of the circuit Diode-triggered SCR (DTSCR) snapback devices have been attached at the corresponding virtual ground nodes. The same type of DTSCR has been also used for the power clamping between the power supply and ground rails. The device triggers at 3.8 V and snaps back to 1.4 V on the positive pulse between RF pad and ground. The DTSCR structure is very similar to the one described in [90]. The equivalent conceptual schematic diagram of the structure is presented in Fig. 6.73. The Vhold diode is used for latch-up immunity during normal operation.

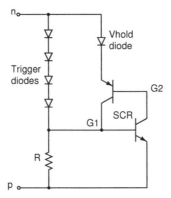

Fig. 6.73 Diode-Triggered SCR schematic diagram.

The choice of the number of trigger diodes poses a trade-off between undesired leak-

6.6 Circuit-Level RF ESD Protection

age and ESD trigger voltage. For lower leakage a higher number of trigger diodes is required, whilst for the reduction of the ESD trigger voltage a lower number of trigger diodes is necessary. Having four diodes is a good compromise for this low-voltage 1.5 V application. The equivalent parasitic capacitance of the DTSCR structure is approximately 0.3 pF.

The concept has been analyzed by electromagnetic simulations in the 2.5D field solver SonnetEM and verified in detail by HBM and TLP measurements.

6.6.2.1 Concept Verification by Circuit Simulation

The full on-chip metallization from the pads to the transistor terminals has been simulated in the field solver SonnetEM for the input and output stage. The structures have been simulated once with the additional port in the middle for attaching an ESD device and once without it, as shown in Fig. 6.74.

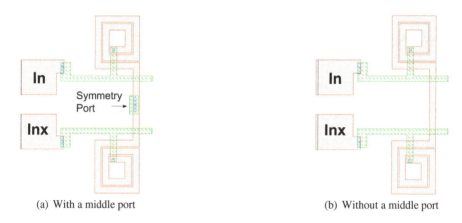

Fig. 6.74 Input stage top level structures simulated in SonnetEM.

The extracted S-parameters have been included in the circuit simulation of the LNA in Agilent ADS, once with the Spice model of the ESD device attached at the node corresponding to the middle port in Fig. 6.74(a) and once without it. The comparison of the simulated and measured small signal forward transmission S-parameter is presented in Fig. 6.75.

As can be observed, the curves for the simulation with the ESD and without the ESD device perfectly overlap. Thus, the high-frequency performance is not affected by attaching the ESD device at the virtual ground.

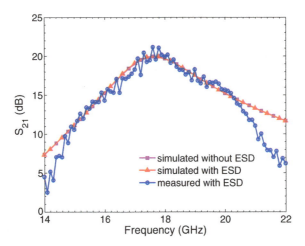

Fig. 6.75 Comparison of measured and simulated LNA gain.

6.6.2.2 Concept Verification by HBM Measurement

The chip has been tested using a HBM simulator according to the ESD standard JEDEC ESD 22A114. The differential RF input and output pads have been verified and the summary of the results is presented in Table 6.7.

Table 6.7 JESD HBM measurements of the LNA RF pins

Path	Positive	Negative
RF input versus GND (kV)	+3	-2.5
RF input versus VDD (kV)	+2.5	-3
RF Inp versus RF Inn (kV)	$> +8$	< -8
RF output versus GND (kV)	+4	-3.5
RF output versus VDD (kV)	+3	-3.5
RF Outp versus RF Outn (kV)	$> +8$	< -8

It can be observed that the RF output exhibits higher pass levels than the RF input for an ESD stress applied between VDD and GND. This is due to the fact that the drain-source path is exposed to the stress at the output, whilst the gate-source path and thus the gate oxide is exposed to the stress at the input. The C11N transistors have a gate oxide as thin as 2.2 nm, as described in section 3.1.2. For the C11N process the typical DC breakdown voltage of the gate oxide is 4.5 V, which is much lower than the junction avalanche breakdown voltage of 7 V.

Additionally, it can be observed from the HBM results in Table 6.7 that stressing RF Inp versus RF Inn provides a very high pass level. Further testing was not possible due to the limits of the available HBM tester. This high level is due to the fact that the small values of the RF inductors introduce a low-impedance path between the RF pins for the incident signals with sub-gigahertz spectral content, as in case of HBM ESD transient events. Thus, the failure is presumably due to the overheating of the transmission lines. Stressing RF Outp versus RF Outn provides a very similar result, due to the above considerations.

6.6.2.3 Concept Verification by TLP Measurement

The chip has been stressed using a TLP pulser with 100 ns pulse length. The TLP measurement [21] has been performed on-wafer using GGB Picoprobe probes and a TDS6124C Tektronix sampling oscilloscope. Rise time filters of 0.59 ns and 10 ns have been used to control the edges of ESD pulses.

The TLP measurement of the path between the RF input and ground of the LNA measured using a 10 ns rise time filter is presented in Fig. 6.76.

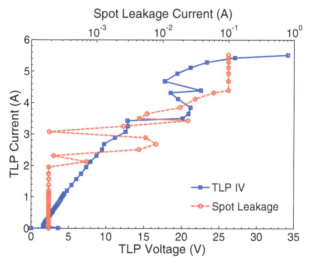

Fig. 6.76 TLP measurement of the path from RF input to ground.

It can be observed that the first failure occurs at 2 A. This corresponds well to the HBM measurement result of +3 kV. According to the simplified rule of thumb, the TLP failure current I_{TLP} measured using pulses with 10 ns rise time is related to the HBM protection level V_{HBM} as follows [91]

$$V_{HBM}(V) \approx I_{TLP}(A) \cdot R_{HBM}(\Omega), \qquad (6.42)$$

where R_{HBM} is the human body resistance of 1.5 kΩ used in the HBM model. From correlation of time-domain waveforms and spot leakage curves it has been observed that the first circuit failure at 2 A occurs when the instantaneous voltage on the RF pad exceeds 7 V. It has been reported in [92] that for the C11N technology the gate oxide breakdown voltage under stress with pulse length of 100 ns is about 6.5 V for NMOSFET and 8.5 V for PMOSFET. This correlation implies that the failure at 2 A in Fig. 6.76 is due to the MOSFET oxide. Additional leakage increase, observed at the TLP current of 3 A, corresponds to the failure of the ESD device. The final failure at 5.3 A occurs due to the overheating and damage of the transmission lines.

Additionally, it has been observed that an ESD stress from RF input to ground using TLP pulses with a rise time of 0.59 ns results in a failure current of 0.2 A, which is much lower compared to the result obtained using the pulses with a rise time of 10 ns. This is due to a voltage overshoot reaching 7 V for the faster pulse in Fig. 6.77, which occurs since the DTSCR snapback structure is too slow to switch on for the fast pulses, as also reported in [93]. Thus, the ESD robustness can be further increased by using a faster ESD device, as e.g. diode. However, a larger area is then needed to achieve the same protection level. The different shapes of the two waveforms in Fig. 6.77, measured at the pads of the LNA, are due to two different rise time filter modules used for the measurements.

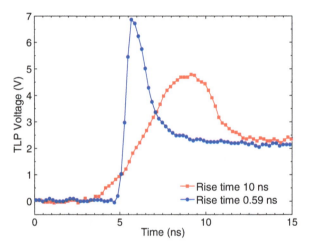

Fig. 6.77 Transient voltages for 0.2 A pulse with 10 ns and 0.59 ns rise times.

The voltage overshoot due to the inductors in the discharge path is small because of their small inductance values. This is due to the fact that at multi-gigahertz frequencies the inductance values, required for tuning out the parasitic capacitances or for inductive peaking, are usually quite small. However, at higher TLP currents and faster rise times the overshoot may become significant.

6.6.3 Transformer Protection Concept

Most of the previously described techniques need semiconductor ESD protection devices that shunt high currents. The discrimination, whether a signal is a wanted signal that occurs during normal circuit operation or an ESD pulse that has to be shunted, is done by the signal level. Once a large input signal level exceeds a certain turn-on voltage, the ESD devices start to conduct a large current, since it is "interpreted" as an ESD event.

However, there is an additional option of providing protection for ESD events between an RF pad and ground. This option does not require a semiconductor ESD device, but implements a transformer, which can also simultaneously be used for input matching and as a balun to convert between single-ended to differential signals. Thus, this approach is advantageous for saving chip area, since several functions are realized in a single structure. The need for on-chip baluns has been addressed in detail in section 6.5. Furthermore, the performance deterioration due to highly capacitive ESD protection devices is avoided. The schematic diagram of the protection principle is presented in Fig. 6.78.

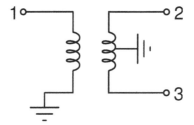

Fig. 6.78 Schematic diagram of the transformer ESD protection.

In this protection approach the discrimination whether it is a wanted signal or an ESD event is done by the spectral content. For high-frequency signals around the center operation frequency, in this case of 24 GHz, the transformer couples the energy into the circuit. For slow ESD events as e.g. HBM with rise times of 10 ns that have a spectral content below 1 GHz, the primary side of the transformer offers a very low impedance path to ground.

A simplified equivalent circuit of a transformer is shown in Fig. 6.79, where L_1 and L_2 are the primary and secondary inductances, the parameter n denotes the number of turns ratio and k is the magnetic coupling coefficient [53]. The shunt inductance kL_1 on the primary side, commonly referred to as the *magnetizing* inductance, describes the portion of the primary inductance that participates in magnetic coupling between the sides. This is also the reason why transformers are not able to operate at low frequencies or need to be very large [17]. At lower frequencies this inductance offers a very low impedance path to ground for the configuration in Fig. 6.78. Thus, the high ESD currents flow directly to ground, whilst only a negligible portion is coupled to the main circuit.

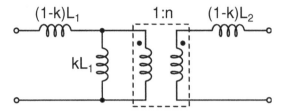

Fig. 6.79 Equivalent circuit of a transformer.

Similarly to the description in section 6.6.2.3, the voltage overshoot at the primary side of the transformer is small because of the small inductance values, typically used at multi-gigahertz frequencies. The series resistance of the primary transformer winding that experiences the ESD pulse, has to be minimized in order to minimize the power dissipation in the structure. Obviously, the amount of energy, that can be conducted until the passive components fail, limits the protection level. The failure levels of passive structures as e.g. pads, vias, transmission lines or inductors in C11N have been studied in detail in [86]. The typical HBM failure levels of inductors with 10 µm wide lines in the 1.35 µm thick top aluminum level are above 8 kV.

The spectral discrimination has also been used in the approach presented in the previous section, with the difference that in that case the ESD events were "detected" and "forwarded" to an ESD clamping device, whilst here the ESD events are "detected" and shunted using the same passive structure.

However, it has to be noted that this approach is only applicable when the harmonics of the ESD events occur at low frequencies. Thus, it will be less efficient if a circuit has to be protected against CDM pulses with rise time of 100 ps.

The concept has been verified in measurement of a 24 GHz cascode LNA in C11N CMOS. The following sections present the test circuit design, its physical realization, high-frequency measurements and TLP testing of the circuit.

6.6 Circuit-Level RF ESD Protection

6.6.3.1 Test LNA Circuit Design

The conceptual circuit diagram of the test LNA is presented in Fig. 6.80.

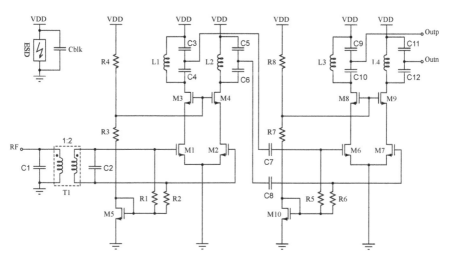

Fig. 6.80 Schematic diagram of the cascode LNA in CMOS.

It is a differential two-stage cascode LNA with parallel tanks as load. The stages are AC-coupled using capacitors C7 and C8. The topology is very similar to the one described in section 6.1.2 for the SiGe LNA. Due to the low transistor transconductance, two stages have been used. Similarly as for the cascode LNA in SiGe, the impedance transformation is realized using a tapped capacitance tank circuit. The source degeneration inductors have been omitted in this case, as for the RF transconductance stage of the active CMOS mixer in section 6.2.1.1. It was possible in this circuit, since the parasitic capacitance of the transistors M1, M2 along with C2 resonates with the secondary side of the transformer at 24 GHz and since the real part of the impedance is dominated by the gate resistance. Omitting the source degeneration inductors offers considerable area saving and has minimal impact on matching. The bias network for both the amplification and cascode transistors are also realized similarly to description in section 6.2.1.1.

In the proposed realization in Fig. 6.80, the transformer is tuned by capacitors C1 and C2 to resonate at 24 GHz. At the single-ended port it offers a good matching to 50 Ω. The capacitor C2 is also used to set the phase relation at the unbalanced side such that there is 180° between the ports 2 and 3. Obviously, this tuned transformer has a narrow bandwidth.

This circuit is only used as a proof of concept to demonstrate the ESD protection approach, thus the RF performance has not been optimized for a particular parameter, as in previous sections.

6.6.3.2 Test LNA Realization and Measurement

The circuit has been processed in Infineon's C11N technology. The annotated chip micrograph is presented in Fig. 6.81. The size including the pads is 0.28 mm^2.

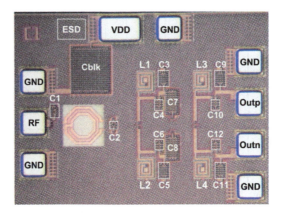

Fig. 6.81 Chip micrograph of the test LNA circuit (0.6 mm × 0.47 mm).

The 1:2 transformer, used in this circuit, is described in detail in section 5.1.2.2. In the physical realization, the transformer has a very low DC series resistance of 0.6 Ω. This is advantageous for efficient ESD protection, as explained previously.

The chip was thinned to 185 μm and mounted on a board. The LNA consumes 17 mA from a single 1.5 V supply. The measured gain and NF of the LNA after de-embedding are presented in Fig. 6.82. The LNA achieves a gain of 7.5 dB and an NF of 6.2 dB around 24 GHz.

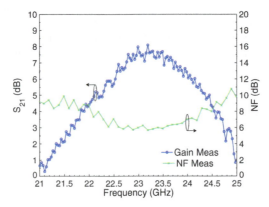

Fig. 6.82 Measured gain and noise figure of cascode LNA.

6.6 Circuit-Level RF ESD Protection

The measured port matching of the LNA is presented in Fig. 6.83. A good port matching of better than 15 dB is achieved around the frequency of 24 GHz.

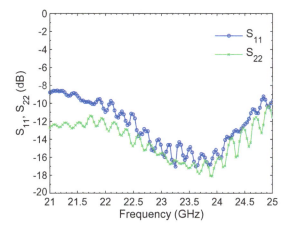

Fig. 6.83 Measured input matching of cascode LNA.

6.6.3.3 Concept Verification by TLP Measurement

The ESD robustness has been analyzed using TLP stress pulses having 10 ns rise time and 100 ns width applied between the RF input to ground. The obtained I/V curve is presented in Fig. 6.84.

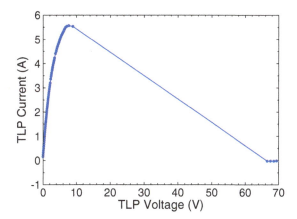

Fig. 6.84 TLP I/V characterization of the path RF input to ground.

The protection fails at a current level of 5.5 A, which corresponds to an HBM hardness above 8 kΩ. To author's knowledge this is the highest reported to date HBM hardness for an RF circuit in this frequency range.

It is interesting to note in Fig. 6.84 that when the protection fails it becomes "open". This means that the failure occurs due to damage of transmission lines that are not able to handle this high current. This shows that with an increasing TLP current level the transformer fails before the MIM capacitor. This is partially due to very low series resistance of the lines, due to slow rise time and sufficiently high sustainable transient voltage on MIM capacitors of about 40 V.

Therefore, this approach is applicable to achieve high protection levels at multigigahertz frequency range against ESD events with slow rise times.

References

1. J. E. Lilienfeld, *Device for controlling electric current*, United States Patent number 1900018, 1928.
2. J. Bardeen and W. H. Brattain, "The Transistor, A Semiconductor Triode", *Physical Review*, vol. 74, pp. 230--231, July 1948.
3. H. Darabi and J. Chiu, "A Noise Cancellation Technique in Active RF-CMOS Mixers", *IEEE Journal of Solid-State Circuits*, vol. 40, pp. 2628--2632, Dec 2005.
4. J. Koh, D. Schmitt-Landsiedel, R. Thewes, and R. Brederlow, "A Complementary Switched MOSFET Architecture for the 1/f Noise Reduction in Linear Analog CMOS ICs", *IEEE Journal of Solid-State Circuits*, vol. 42, pp. 1352--1361, June 2007.
5. S. Maas, "A GaAs MESFET Mixer with Very Low Intermodulation", *IEEE Transactions on Microwave Theory and Techniques*, vol. 35, pp. 425--429, April 1987.
6. S. Nguyen, "CMOS low-noise amplifier design optimization techniques", *IEEE Transactions on Microwave Theory and Techniques*, vol. 52, pp. 1433--1442, May 2004.
7. S. T. Nicolson and S. P. Voinigescu, "Methodology for Simultaneous Noise and Impedance Matching in W-Band LNAs", *in IEEE Compound Semiconductor IC Symposium (CSICS) Digest*, pp. 279--282, San Antonio, USA, November 2006.
8. V. Issakov, M. Tiebout, Y. Cao, A. Thiede, and W. Simbürger, "A low power 24 GHz LNA in 0.13 μm CMOS", *in IEEE Conference on Microwaves, Communications, Antennas and Electronic Systems (COMCAS)*, pp. 1--10, Tel Aviv, Israel, May 2008.
9. V. Issakov, H. Knapp, M. Wojnowski, A. Thiede, W. Simbürger, G. Haider, and L. Maurer, "ESD-protected 24 GHz LNA for Radar Applications in SiGe:C Technology", *in Topical Meeting on Silicon Monolithic Integrated Circuits in RF Systems (SiRF)*, pp. 1--4, San Diego, USA, January 2009.
10. D. K. Shaeffer and T. H. Lee, "A 1.5-V, 1.5-GHz CMOS Low Noise Amplifier", *IEEE Journal of Solid-State Circuits*, vol. 32, pp. 745--759, May 1997.
11. M. J. Deen, C.-H. Chen, S. Asgaran, G. A. Rezvani, J. Tao, and Y. Kiyota, "High-Frequency Noise of Modern MOSFETs: Compact Modeling and Measurement Issues", *IEEE Transactions on Electron Devices*, vol. 53, pp. 2062--2081, September 2006.
12. G. Dambrine, H. Happy, F. Danneville, and A. Cappy, "A New Method for On Wafer Noise Measurement", *IEEE Transactions on Microwave Theory and Techniques*, vol. 41, pp. 375--381, March 1993.
13. T. O. Dickson, K. H. K. Yau, T. Chalvatzis, A. M. Mangan, E. Laskin, R. Beerkens, P. Westergaard, M. Tazlauanu, M.-T. Yang, and S. P. Voinigescu, "The Invariance of Characteristic Current Densities in Nanoscale MOSFETs and Its Impact on Algorithmic Design Methodologies and Design Porting of Si(Ge) (Bi)CMOS High-Speed Building Blocks", *IEEE Journal of Solid-State Circuits*, vol. 41, pp. 1830--1845, August 2006.

References 159

14. A. S. Sedra and K. C. Smith, *Microelectronic Circuits*, Oxford University Press, 2004.
15. D. Zoschg, W. Wilhelm, T. F. Meister, H. Knapp, H.-D. Wohlmuth, K. Aufinger, M. Wurzer, J. Bock, H. Schafer, and A. Scholtz, "2 dB noise figure, 10.5 GHz LNA using SiGe bipolar technology", *Electronics Letters*, vol. 35, pp. 2195--2196, December 1999.
16. E. van der Heijden, H. Veenstra, D. Hartskeerl, M. Notten, and D. van Goor, "Low Noise Amplifier with integrated balun for 24 GHz car radar", *in Topical Meeting on Silicon Monolithic Integrated Circuits in RF Systems (SiRF)*, pp. 78--81, Orlando, USA, January 2008.
17. T. H. Lee, *The Design of CMOS Radio Frequency Integrated Circuits*, Cambridge University Press, 1998.
18. R. Welch and *et al.*, "The Effects of Feedback Capacitance on Thermally Shunted Heterojunction Bipolar Transistor's Linearity", *in International Conference on Compound Semiconductor Manufacturing Technology (Mantec)*, Vancouver, Canada, May 1999, available online at http://www.csmantech.org/Digests/1999/PDF/41.pdf.
19. G. Niu, "Noise in SiGe HBT RF Technology: Physics, Modeling, and Circuit Implications", *Proc. of the IEEE*, vol. 93, pp. 1583--1597, September 2005.
20. S. P. Voinigescu, M. C. Maliepaard, J. L. Showell, G. E. Babcock, D. Marchesan, M. Schroter, P. Schvan, and D. L. Harame, "A Scalable High-Frequency Noise Model for Bipolar Transistors with Application to Optimal Transistor Sizing for Low-Noise Amplifier Design", *IEEE Journal of Solid-State Circuits*, vol. 32, pp. 1430--1439, September 1997.
21. H. Hyatt, J. Harris, A. Alanzo, and P. Bellew, "TLP measurements for verification of ESD protection device response", *in Electrical Overstress/ Electrostatic Discharge (EOSESD) Symposium*, pp. 111--120, Anaheim, USA, September 2000.
22. K.-W. Yu and M. F. Chang, "CMOS K-band LNAs design counting both interconnect transmission line and RF pad parasitics", *in IEEE Radio Frequency Integrated Circuits (RFIC) Symposium Digest*, pp. 101--104, Fort Worth, USA, June 2004.
23. B. A. Floyd, L. Shi, Y. Taur, I. Lagnado, and K. K. O, "A 23.8-GHz SOI CMOS Tuned Amplifier", *IEEE Transactions on Microwave Theory and Techniques*, vol. 50, pp. 2193--2196, September 2002.
24. S. Pruvost, L. Moquillon, E. Imbs, M. Marchetti, and P. Garcia, "Low Noise Low Cost Rx Solutions for Pulsed 24 GHz Automotive Radar Sensors", *in IEEE Radio Frequency Integrated Circuits (RFIC) Symposium Digest*, pp. 387--390, Honolulu, Hawaii, June 2007.
25. B. Gilbert, "A precise four-quadrant multiplier with subnanosecond response", *IEEE Journal of Solid-State Circuits*, vol. 3, pp. 365--373, December 1968.
26. S. A. Maas, *Nonlinear Microwave and RF Circuits*, Artech House, 2003.
27. M. Hossain, B. M. Frank, and Y. M. Antar, "Performance of a low voltage highly linear 24 GHz down conversion mixer in 0.18 μm CMOS", *in Topical Meeting on Silicon Monolithic Integrated Circuits in RF Systems (SiRF)*, San Diego, USA, January 2006.
28. N. Shiramizu, T. Masuda, T. Nakamura, and K. Washio, "24-GHz 1-V pseudo-stacked mixer with gain-boosting technique", *in Proc. of European Solid-State Circuits Conference (ESSCIRC)*, pp. 102--105, Edinburgh, UK, September 2008.
29. J. Park, C.-H. Lee, B.-S. Kim, and J. Laskar, "Design and Analysis of Low Flicker-Noise CMOS Mixers for Direct-Conversion Receivers", *IEEE Transactions on Microwave Theory and Techniques*, vol. 54, pp. 4372--4380, December 2006.
30. I. Gresham and A. Jenkins, "A low-noise broadband SiGe mixer for 24 GHz ultra-wideband automotive applications", *in Proc. Radio and Wireless Symposium (RAWCON)*, pp. 361--364, Boston, USA, August 2003.
31. R. M. Kodkani and L. E. Larson, "A 24-GHz CMOS Passive Subharmonic Mixer/Downconverter for Zero-IF Applications", *IEEE Transactions on Microwave Theory and Techniques*, vol. 56, pp. 1247--1256, May 2008.
32. T. Chang and J. Lin, "1-11 GHz Ultra-Wideband Resistive Ring Mixer in 0.18 μm CMOS Technology", *in IEEE Radio Frequency Integrated Circuits (RFIC) Symposium Digest*, San Francisco, USA, June 2006.
33. N. Kim, V. Aparin, and L. E. Larson, "A Resistively Degenerated Wide-Band Passive Mixer with Low Noise Figure and +60 dBm IIP2 in 0.18 μm CMOS", *in IEEE Radio Frequency Integrated Circuits (RFIC) Symposium Digest*, pp. 185--188, Atlanta, USA, June 2008.

34. V. Issakov, H. Knapp, M. Tiebout, A. Thiede, W. Simbürger, and L. Maurer, "Comparison of 24 GHz Low-Noise Mixers in CMOS and SiGe:C Technologies", *in European Microwave Integrated Circuits Conference (EuMIC)*, pp. 184--187, Rome, Italy, October 2009.
35. V. Issakov, A. Thiede, L. Verweyen, and L. Maurer, "Wideband Resistive Ring Mixer for Automotive and Industrial Applications in 0.13 μm CMOS", *in German Microwave Conference (GeMiC)*, pp. 1--4, Munich, Germany, March 2009.
36. V. Issakov, H. Knapp, A. Thiede, W. Simbürger, and L. Maurer, "A $22 - 38$ GHz Integrated Passive Mixer in SiGe:C Technology", *in submitted to IEEE MTT-S International Microwave (IMS) Symposium*, Anaheim, USA, June 2010.
37. S. P. Voinigescu, T. O. Dickson, M. Gordon, C. Lee, T. Yao, A. Mangan, K. Tang, and K. Yau, *Si- based Semiconductor Components for Radio-Frequency Integrated Circuits (RF IC)*, chapter RF and Millimeter-Wave IC Design in the Nano-(Bi)CMOS Era, pp. 33--62, Transworld Research Network, 2006.
38. N. Zhang, H. Xu, H.-T. Wu, and K.-K. O, "W-Band Active Down-Conversion Mixer in Bulk CMOS", *IEEE Microwave and Wireless Components Letters*, vol. 19, pp. 98--100, February 2009.
39. F. Ellinger, *Radio Frequency Integrated Circuits and Technologies*, chapter 4, Springer, 2007.
40. C.P. Yue and S.S. Wong, "On-chip spiral inductors with patterned ground shields for Si-based RF ICs", *IEEE Journal of Solid-State Circuits*, vol. 33, pp. 743--752, May 1998.
41. J. Rogers and C. Plett, *Radio Frequency Integrated Circuit Design*, chapter 5, Artech House, 2003.
42. F. Ellinger, "26-34 GHz CMOS mixer", *Electronics Letters*, vol. 40, pp. 1417--1419, October 2004.
43. C.-S. Lin, P.-S. Wu, H.-Y. Chang, and H. Wang, "A 9-50-GHz Gilbert-cell down-conversion mixer in 0.13-um CMOS technology", *IEEE Microwave and Wireless Components Letters*, vol. 16, 2006.
44. A. A. M. Saleh, *Theory of resistive mixers*, MIT Press, 1971.
45. A. J. Kelly, "Fundamental Limits on Conversion Loss of Double Sideband Resistive Mixers", *IEEE Transactions on Microwave Theory and Techniques*, vol. 25, pp. 867--869, November 1977.
46. E. W. Lin and W. H. Ku, "Device considerations and modeling for the design of an InP-based MODFET millimeter-wave resistive mixer with superior conversion efficiency", *IEEE Transactions on Microwave Theory and Techniques*, vol. 43, pp. 1951--1959, August 2001.
47. F. Ellinger, "26.5-30-GHz Resistive Mixer in 90-nm VLSI SOI CMOS Technology With High Linearity for WLAN", *IEEE Transactions on Microwave Theory and Techniques*, vol. 53, pp. 2559--2565, August 2005.
48. V. Issakov, A. Thiede, L. Verweyen, and M. Tiebout, "0.5-25 GHz inductorless single-ended resistive mixer in 0.13 μm CMOS", *Electronics Letters*, vol. 45, pp. 108--109, January 2009.
49. S. Sankaran and K. K. O, "Schottky diode with cutoff frequency of 400 GHz fabricated in 0.18 μm CMOS", *Electronics Letters*, vol. 41, pp. 506--508, April 2005.
50. L. Roselli, F. Alimenti, M. Comez, V. Palazzari, F. Placentino, N. Porzi, and A. Scarponi, "A cost driven 24 GHz Doppler radar sensor development for automotive applications", *in European Radar Conference (EuRAD)*, pp. 335--338, Paris, France, October 2005.
51. Y. H. Lin and Y. J. Chan, "2.4 GHz single balanced diode mixer fabricated on Al_2O_3 substrate", *in Proc. Asia-Pacific Microwave Conference (APMC)*, pp. 218--221, Singapore, November 1999.
52. L. Verweyen, H. Massler, M. Neumann, U. Schaper, and W. H. Haydl, "Coplanar integrated mixers for 77-GHz automotive applications", *IEEE Microwave and Guided Wave Letters*, vol. 8, pp. 38--40, January 1998.
53. I. J. Bahl, *Lumped elements for RF and microwave circuits*, Artech House, 2003.
54. P.-S. Wu, C.-S. Lin, T.-W. Huang, H. Wang, Y.-C. Wang, and C.-S. Wu, "A millimeter-wave ultra-compact broadband diode mixer using modified Marchand balun", *in Gallium Arsenide and Other Semiconductor Application Symposium (EGAAS)*, pp. 349--352, Paris, France, October 2005.

References 161

55. X. Guan and A. Hajimiri, "A 24-GHz CMOS front-end", *IEEE Journal of Solid-State Circuits*, vol. 39, pp. 368--373, February 2004.
56. V. Geffroy, G. De Astis, and E. Bergeault, "RF mixers using standard digital CMOS 0.35 μm process", *in IEEE MTT-S International Microwave Symposium (IMS) Digest*, pp. 83--86, Phoenix, USA, May 2001.
57. M. Voltti, T. Koivi, and E. Tiiliharju, "Comparison of active and passive mixers", *in Proc. European Conference on Circuit Theory and Design (ECCTD)*, pp. 890--893, Sevilla, Spain, August 2007.
58. V. Issakov, D. Šiprak, M. Tiebout, A. Thiede, W. Simbürger, and L. Maurer, "Comparison of 24 GHz Receiver Front-Ends using Active and Passive Mixers in CMOS", *IET Circuits, Devices & Systems*, vol. 3, pp. 340--349, December 2009.
59. Y. Tsividis, *Operation and Modeling of the MOS Transistor*, McGraw-Hill, 2nd edition, 1999.
60. B. Dehlink, H.-D. Wohlmuth, K. Aufinger, T. F. Meister, J. Böck, and A. L. Scholz, "A low-noise amplifier at 77 GHz in SiGe:C bipolar technology", *in IEEE Compound Semiconductor IC Symposium (CSICS) Digest*, pp. 287--290, Palm Springs, USA, November 2005.
61. K. Chang, *RF and Microwave Wireless Systems*, Wiley, 2000.
62. W. Sansen, "Distortion in elementary transistor circuits", *IEEE Transactions on Circuits and Systems II: Analog and Digital Signal Processing*, vol. 46, pp. 315--325, March 1999.
63. R. J. Baker, H. W. Li, and D. E. Boyce, *CMOS Circuit Design, Layout, and Simulation*, IEEE Press, 1998.
64. Y.-H. Chen, H.-H. Hsieh, and L.-H. Hsieh, "A 24-GHz Receiver Frontend With an LO Signal Generator in 0.18-μm CMOS", *IEEE Transactions on Microwave Theory and Techniques*, vol. 56, pp. 1043--1051, May 2008.
65. M. Törmänen and H. Sjöland, "Two 24 GHz Receiver Front-ends in 130 nm CMOS using SOP Technology", *in IEEE Radio Frequency Integrated Circuits (RFIC) Symposium Digest*, pp. 559--562, Boston, USA, June 2009.
66. T. Yu and G. M. Rebeiz, "A 24 GHz 4-channel phased-array receiver in 0.13 μm CMOS", *in IEEE Radio Frequency Integrated Circuits (RFIC) Symposium Digest*, pp. 361--364, Atlanta, USA, June 2008.
67. H. Veenstra, E. van der Heijden, M. Notten, and G. Dolmans, "A SiGe-BiCMOS UWB Receiver for 24 GHz Short-Range Automotive Radar Applications", *in IEEE MTT-S International Microwave Symposium (IMS) Digest*, pp. 1791--1794, Honolulu, Hawaii, June 2007.
68. S.-Y. Kim, K. V. Buer, E. Imbs, and G. M. Rebeiz, "An 18-20 GHz Subharmonic Satellite Down-Converter in 0.18 μm SiGe Technology", *in Topical Meeting on Silicon Monolithic Integrated Circuits in RF Systems (SiRF)*, pp. 1--4, San Diego, USA, January 2009.
69. V. Issakov, K. L. R. Mertens, M. Tiebout, A. Thiede, and W. Simbürger, "Compact Quadrature Receiver for 24 GHz Radar Applications in 0.13 μm CMOS", *Electronics Letters*, vol. 46, no.1, pp. 315--325, January 2010.
70. V. Issakov, H. Knapp, F. Magrini, A. Thiede, W. Simbürger, and L. Maurer, "Low-Noise ESD-protected 24 GHz Receiver for Radar Applications in SiGe:C Technology", *in Proc. of European Solid-State Circuits Conference (ESSCIRC)*, pp. 308--311, Athens, Greece, September 2009.
71. M. G. M. Notten and H. Veenstra, "60 GHz quadrature signal generation with a single phase VCO and polyphase filter in a 0.25 μm SiGe BiCMOS technology", *in Proc. Bipolar / BiCMOS Circuits and Technology Meeting (BCTM)*, pp. 178--181, Monterey, USA, October 2008.
72. B. Gilbert, "Fundamental aspects of modern active mixer design", *in IEEE International Solid-State Circuits Conference (ISSCC)*, San Francisco, February 2000. IEEE, Short Course "Circuits and Devices for RF Wireless Networks".
73. A. Y.-K. Chen, H.-B. Liang, Y. Baeyens, Y.-K. Chen, J. Lin, and Y.-S. Lin, "Wideband Mixed Lumped-Distributed-Element 90° and 180° Power Splitters on Silicon Substrate for Millimeter-Wave Applications", *in IEEE Radio Frequency Integrated Circuits (RFIC) Symposium Digest*, pp. 449--452, Atlanta, USA, June 2008.

74. W. Bakalski, W. Simbürger, H. Knapp, H.-D. Wohlmuth, and A. L. Scholtz, "Lumped and Distributed Lattice-type LC-Baluns", *in IEEE MTT-S International Microwave Symposium (IMS) Digest*, pp. 209--212, Seattle, USA, June 2002.
75. R. C. Frye, S. Kapur, and R. C. Melville, "A 2-GHz Quadrature Hybrid Implemented in CMOS Technology", *IEEE Journal of Solid-State Circuits*, vol. 38, pp. 550--555, March 2003.
76. S. J. Parisi, "A Lumped Element Rat Race Coupler", *Applied Microwave*, vol. , pp. 8493, September 1989.
77. Y. Leblebici, *ESD Protection and Reliability*, Advanced Engineering Course on "Design for hostile environment: Automotive and Industrial", mead education edition, 2008.
78. A. Amerasekera and C. Duvvury, *ESD in Silicon Integrated Circuits*, Wiley, 2002.
79. S. Cao, T. W. Chen, S. G. Beebe, and R. W. Dutton, "ESD design challenges and strategies in deeply scaled integrated circuits", *in Custom Integrated Circuits Conference (CICC)*, pp. 681--688, San Jose, USA, September 2009.
80. W. Soldner, M. Streibl, U. Hodel, M. Tiebout, H. Gossner, D. Schmitt-Landsiedel, J. H. Chun, C. Ito, and R. W. Dutton, "RF ESD Protection Strategies: Codesign vs. low-C protection", *in Electrical Overstress/ Electrostatic Discharge (EOSESD) Symposium*, pp. 33--42, Tucson, USA, September 2005.
81. V. Vassilev, S. Thijs, P. L. Segura, P. Leroux, P. Wambacq, G. Groeseneken, M. I. Natarajan, M. Steyaert, and H. E. Maes, "Co-design methodology to provide high ESD protection levels in the advanced RF circuits", *in Electrical Overstress/ Electrostatic Discharge (EOSESD) Symposium*, pp. 1--9, Las Vegas, USA, September 2003.
82. M.-D. Ker and C.-M. Lee, "ESD Protection Design for Giga-Hz RF CMOS LNA with Novel Impedance-Isolation Technique", *in Electrical Overstress/ Electrostatic Discharge (EOSESD) Symposium*, pp. 1--10, Las Vegas, USA, September 2003.
83. B. Kleveland, T. J. Maloney, I. Morgan, L. Madden, T. H. Lee, and S. S. Wong, "Distributed ESD protection for high-speed integrated circuits", *IEEE Electron Device Letters*, vol. 21, pp. 390--392, August 2000.
84. J. Shorb, X. Li, and D. J. Allstot, "A resonant pad for ESD protected narrowband CMOS RF applications", *in Proc. on International Symposium on Circuits and Systems (ISCAS)*, pp. I--61 -- I--64, Bangkok, Thailand, June 2003.
85. S. Galal and B. Razavi, "Broadband ESD protection circuits in CMOS technology", *in IEEE International Solid-State Circuits Conference (ISSCC)*, pp. 182--486, San Francisco, February 2003. IEEE.
86. V. Issakov, D. Johnsson, Y. Cao, M. Tiebout, M. Mayerhofer, W. Simbürger, and L. Maurer, "ESD Concept for High-Frequency Circuits", *in Electrical Overstress/ Electrostatic Discharge (EOSESD) Symposium*, pp. 221--227, Tucson, USA, September 2008.
87. D. Linten, M. I. Natarajan, S. Thijs, S. Van Huylenbroeck, S. Xiao, G. Carchon, S. Decoutere, M. Sawada, T. Hasebe, and G. Groeseneken, "Implementation of 6 kV ESD Protection for a 17 GHz LNA in 130 nm SiGeC BiCMOS", *in Proc. on International Conference on Semiconductor Electronics (ICSE)*, pp. A7 -- A12, Kuala Lumpur, Malaysia, July 2006.
88. J. Borremans, S. Thijs, P. Wambacq, D. Linten, Y. Rolain, and M. Kuijk, "A 5 kV HBM transformer-based ESD protected 5-6 GHz LNA", *in IEEE Symposium on VLSI Circuits*, pp. 100--101, Kyoto, Japan, June 2007.
89. Y. Cao, V. Issakov, and M. Tiebout, "A 2 kV ESD protected 18 GHz LNA with 4 dB NF in 0.13 μm CMOS", *in IEEE International Solid-State Circuits Conference (ISSCC)*, pp. 194--606, San Francisco, February 2008. IEEE.
90. M. P. J. Mergens, C. C. Russ, K. G. Verhaege, J. Armer, P. C. Jozwiak, R. P. Mohn, B. Keppens, and C. S. Trinh, "Speed optimized diode-triggered SCR (DTSCR) for RF ESD protection of ultra-sensitive IC nodes in advanced technologies", *IEEE Transactions on Device and Materials Reliability*, vol. 5, pp. 532--542, September 2005.
91. A. Z. H. Wang, *On-Chip ESD Protection for Integrated Circuits: An IC Design Perspective*, Kluwer, 2002.

References 163

92. A. Ille, W. Stadler, T. Pompl, H. Gossner, T. Brodbeck, K. Esmark, P. Riess, D. Alvarez, K. Chatty, and R. Gauthier and A. Bravaix, ''Reliability aspects of gate oxide under ESD pulse stress'', *in Electrical Overstress/ Electrostatic Discharge (EOSESD) Symposium*, pp. 1--10, Anaheim, USA, September 2007.
93. H. Gossner, ''ESD protection for the deep sub micron regime - a challenge for design methodology'', *in Proc. VLSI Design*, pp. 809--818, Mumbai, India, January 2004.

Chapter 7
Radar Transceiver Circuits

In the previous chapter the design of receivers in CMOS and SiGe technologies has been presented. As explained there, the main advantage of CMOS is the possibility of highest integration with analog and digital circuitry. High integration may contribute to a considerable cost reduction of the module by reducing the amount of required discrete components and thus the BOM costs, as well as to a much lower module assembly and functionality verification costs. Therefore, the further circuit development towards a highly-integrated low-cost radar module is considered here in CMOS technology. This chapter describes the integration of the receiver and transmitter paths into a single 24 GHz transceiver chip.

Several publications report integrated 24 GHz receivers [1], [2] or transmitters [3], [4] in CMOS. However, only few publications report integrated transceivers at 24 GHz in a standard digital CMOS technology [5]. Section 7.1 presents a transceiver front-end in Infineon's C11N CMOS technology developed in this work. The circuit comprises an IQ receiver described in section 6.4.1.1 and a transmitter, which is based on the building blocks presented in [6].

In a radar system, the transceiver circuit is integrated with antennas to send and receive signals. Numerous available commercial systems use a single antenna for both transmission and reception and thus are referred to as *monostatic* radars, as explained in detail in section 2.1. However, this requires a duplexer component to provide isolation between transmitter and receiver. This is typically realized using expensive transmit/receive (T/R) switches or circulators. Section 7.2 presents a novel concept developed in this work for compact monostatic radar realization without a duplexer. It is based on merging the power amplifier and mixer in a single circuit. The LNA and the mixer circuit blocks are omitted and the mixer functionality is realized at the output stage of the power amplifier. This allows chip area reduction and considerable cost saving due to omitting the expensive duplexer component. This approach is suitable for applications with relaxed receiver noise figure requirements. Furthermore, section 7.2 presents an example implementing this transceiver principle on circuit level.

V. Issakov, *Microwave Circuits for 24 GHz Automotive Radar in Silicon-based Technologies*, DOI 10.1007/978-3-642-13598-9_7, © Springer-Verlag Berlin Heidelberg 2010
165

7.1 IQ Transceiver in CMOS

This section presents an integrated 24 GHz IQ transceiver in CMOS designed in this work and described in detail in [7]. The circuit includes an LNA, two mixers, an on-chip In-phase and Quadrature (IQ) phase generation, a voltage-controlled oscillator (VCO), a power amplifier (PA) driver and a frequency divider. The circuit is optimized for the lowest chip area and current consumption.

Section 7.1.1 presents a block diagram of the transceiver and describes the design of the building blocks. Section 7.1.2 describes the measurement results of the circuit. Finally, section 7.1.3 summarizes the results and compares the performance with the literature.

7.1.1 IQ Transceiver Circuit Design

The block diagram of the implemented transceiver is presented in Fig. 7.1. The transceiver integrates a quadrature receiver and a transmitter. The power at the output of the PA driver is equally split between the TX output towards the transmit antenna and the local-oscillator (LO) input of the receiver.

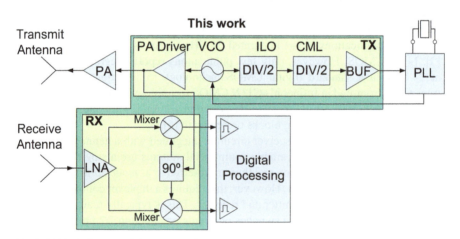

Fig. 7.1 Block diagram of the IQ transceiver.

The IQ receiver block consists of an LNA, two mixers and quadrature generation. A stand-alone receiver is described in detail in section 6.4.1.1. The receiver path has been optimized for lowest noise figure, in order to improve the dynamic range.

Several system-relevant aspects have to be considered during the design of the transmitter. As mentioned in section 2.5, the required parameters of a radar system are influenced by the frequency regulations. As described in Fig. 2.5, the maximal

7.1 IQ Transceiver in CMOS

allowed EIRP transmitter power level in the ISM band at 24 GHz is constrained in the European Union to 20 dBm. Assuming the gain of an antenna array to be typically about 20 dBi, the transmitter output power shall not exceed 0 dBm in the ISM frequency range from 24.05 GHz to 24.25 GHz.

Furthermore, due to the narrow available bandwidth of 200 MHz the realization of a pulse radar is not feasible. Thus, the radar sensors in the ISM band are implemented using the frequency modulated continuous wave principle, described in detail in section 2.3.2. This relaxes the requirement on circuit implementation of the power amplifier, as only phase needs to be modulated and the power amplifier is designed to provide a single high output power level.

The building blocks of the transmitter path have been implemented similarly to those described in [6]. The oscillator frequency is divided by four down to 6 GHz in two stages. The first division by two has been implemented as a differential direct injection-locked oscillator (ILO), since it offers the smallest possible input load to the VCO. The second division by two is implemented as a static current-mode logic (CML) divider. The VCO signal is split and forwarded to the PA driver and ILO without a buffer. This is advantageous for the power consumption, but also allows to profit by the high voltage swing of the VCO.

The conceptual schematic diagram of the VCO and ILO is presented in Fig. 7.2.

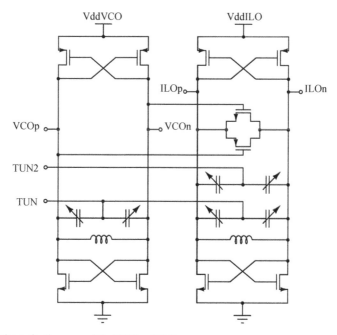

Fig. 7.2 Schematic diagrams of the VCO and ILO.

The VCO has been implemented with a current re-using N-/PMOSFET topology, similar to the one described in [8]. The tank inductance value poses a trade-off between power consumption, tuning range and stability of oscillation. The matching between the LC-tanks of the ILO and of the VCO has been carefully taken into consideration. Both tanks use identical inductors and are laid out using the same varactor and MOS cells. To track the locking range of the ILO to the VCO, both stages share the same tuning input TUN. Additionally, another varactor controlled by the input TUN2 has been added for fine tuning of the ILO. Even though the output of the VCO is loaded by the PA driver and ILO, the circuit is capable of providing a sufficiently wide tuning range.

The next division by two is realized using a classical Master-Slave CML divider topology, described in detail in [9]. The 6 GHz output signal is easier to handle and can be used to control the VCO using an external PLL.

The three-stage PA driver, presented in Fig. 7.3, is based on a circuit topology similar to that of the LNA, described in section 6.1.1.

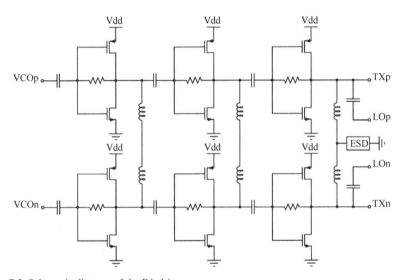

Fig. 7.3 Schematic diagram of the PA driver.

For small signals the N-/PMOSFET topology offers efficient current reuse, whilst for large signals it acts as a digital inverter. Inductors between the stages are used to tune out the parasitic capacitance of the following stage. The output impedance is matched by the parallel LC tank, formed by the output inductors and the capacitances of transistors and pads. The VCO output is fed to the PA driver via AC-coupling capacitors. One part of the PA driver output signal is forwarded to the pads, whilst the other part is AC coupled to the polyphase filter and used as LO input for the mixers.

7.1.2 Measurements of Transceiver

The transceiver has been processed in Infineon's C11N technology. Fig. 7.4 shows the annotated chip micrograph. The chip area including pads is only 0.7 mm².

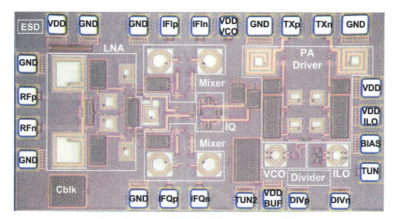

Fig. 7.4 Chip micrograph (1.17 mm × 0.61 mm).

The chip was thinned to 185 μm and mounted on a board for measurements. The transceiver has been characterized using Agilent's Spectrum Analyzer E4448A. The currents of the LNA, two mixers, PA driver, VCO, ILO and CML divider are 12 mA, 4 mA, 23 mA, 4.3 mA, 7.5 mA, and 7.9 mA, respectively, from a 1.5 V supply. Additionally, 24 mA is drawn by the divider output buffer.

The VCO tuning curve is presented in Fig. 7.5. A wide tuning range of 2.6 GHz

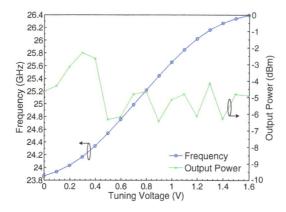

Fig. 7.5 Measured transmitter frequency versus tuning voltage.

has been achieved. The frequency of 24 GHz is achieved for a tuning voltage of 0.18 V. The output power of −3 dBm has been measured in the ISM band. The maximum output power of −2.4 dBm is achieved at the frequency of 24.5 GHz. The transmitter output power is lower than expected due to a lower center frequency of the PA driver and a higher frequency of the VCO.

Fig. 7.6 presents the output spectrum including the cumulative losses of 5.5 dB by the balun, board microstrip lines, connectors, cabling and DC blocking capacitor. The free-running phase noise measurement is presented in Fig. 7.7.

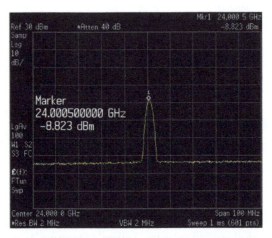

Fig. 7.6 Measured transmitter output spectrum at a tuning voltage of 0.18 V.

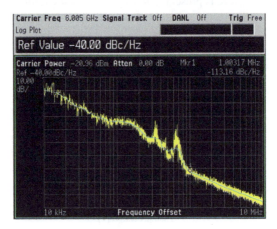

Fig. 7.7 Phase noise measured at the divider output.

7.1 IQ Transceiver in CMOS

The measurement has been performed at the divider output for a VCO frequency of 24 GHz. The phase noise of −113.16 dBc/Hz at the 6 GHz divider output corresponds to at most −101 dBc/Hz at the 24 GHz VCO output.

The receiver path has been measured using a buffer offering a high load impedance at the IF outputs. The receiver conversion gain has been tested by sweeping the VCO frequency over the whole tuning range and adjusting the externally applied RF input frequency to keep the IF at approximately 10 MHz. The RF input power of −40 dBm has been applied to the pads. The noise figure has been measured using the direct noise floor method [10]. The measured conversion gain and noise figure versus RF frequency are presented in Fig. 7.8. As can be seen, the transceiver offers a very low double sideband (DSB) noise figure of 5.5 dB and a sufficient gain of 12 dB over the ISM band. The gain remains above 9 dB and noise figure below 7 dB over the entire frequency range of the VCO.

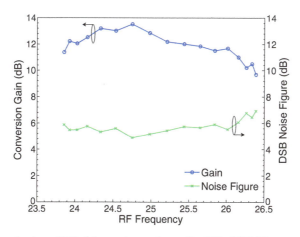

Fig. 7.8 Measured gain and NF of the receiver path or a fixed IF of 10 MHz.

The input-referred 1 dB compression point is measured for an RF frequency of 24.01 GHz and an LO frequency of 24 GHz to be −16.2 dBm, as presented in Fig. 7.9. This linearity is sufficient for radar receiver front-end applications, allowing the sensor to operate properly under a blocker case, i.e. in presence of a strong nearby target.

7.1.3 Results Summary and Comparison

A compact 24 GHz low-power CMOS transceiver for radar applications has been developed in this work. On a record minimal area of 0.7 mm² the circuit integrates an LNA, two mixers, an on-chip quadrature generation, a VCO, a PA driver and a frequency division by four. It consumes only 88 mW from a single 1.5 V supply.

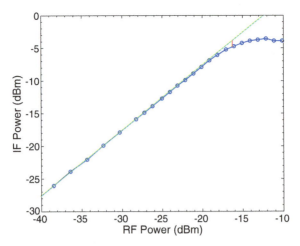

Fig. 7.9 IP1dB at an RF of 24.01 GHz and an IF of 10 MHz.

The measured transceiver performance at 24 GHz is summarized in Table 7.1 and compared with a recently published state-of-the-art transceiver for FMCW radar applications operating in the 24 GHz ISM band. It offers comparable performance and achieves a lower receiver noise figure. The presented transceiver is suitable for low-cost mass-market 24 GHz radar applications in the ISM band.

Table 7.1 Transceiver measured performance summary and comparison

Parameter	[11]	This work
Technology node	SiGe 0.18 μm	CMOS 0.13 μm
f_T (GHz)	150	100
Receiver gain (dB)	18	12
Receiver DSB NF (dB)	10	5.5
Receiver IP1dB (dBm)	-15	-16.2
Transmitter power (dBm)	7	-3
Transmitter tuning range (GHz)	NA	23.8-26.4
Transmitter phase noise (dBc/Hz)	-82 @ 100 kHz[1]	-101 @ 1 MHz[2]
P_{dc} (mW)	275[3]	88
V_{dc} (V)	3.5	1.5
Size (mm^2)	NA	0.7

[1] The VCO is locked using PLL.
[2] Free running VCO.
[3] Very high level integration.

7.2 Merged Power-Amplifier-Mixer Transceiver

This section presents a novel radar transceiver architecture that avoids using a bulky and expensive duplexer, such as e.g. T/R switch or circulator, in a monostatic radar implementation. This principle is verified in measurement of a merged power-amplifier-mixer circuit designed in C11N CMOS technology and presented in [12]. The receive path of the test circuit has been optimized for low conversion loss and the power amplifier for high output power.

Section 7.2.1 discusses the system-relevant considerations of the proposed architecture. Section 7.2.2 presents the circuit-level implementation of the proposed concept. Section 7.2.3 describes the achieved measured performance of the test circuit. Finally, section 7.2.4 summarizes the measurement results and compares them with the state-of-the-art power amplifiers and mixers.

7.2.1 System Considerations

The simplistic block diagram of a classical monostatic radar architecture is presented in Fig. 7.10(a), where f_0 is the transmitted signal and f_d is the down-converted signal. The transmitted and received signals are separated by a duplexer. It is required to provide sufficient isolation to prevent receiver desensitization due to gain drop in the presence of a strong interferer.

For relaxed radar system specification, the architecture can be considerably simplified, as shown in Fig. 7.10(b). The proposed topology omits LNA and mixer blocks and uses the transistors of the PA simultaneously for both amplification and mixing. The down-converted signal at intermediate frequency is obtained by corresponding filters. This solution offers the significant advantage of omitting the duplexer, but has the disadvantage of higher conversion loss and noise figure.

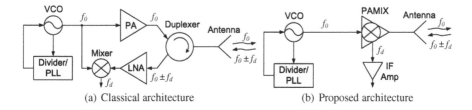

(a) Classical architecture   (b) Proposed architecture

Fig. 7.10 Block diagrams of the monostatic radar architectures.

The noise figure of a receiver consisting of an LNA and a mixer, as shown in Fig. 7.10(a), is given by the Friis formula in Eq. (6.29). Considering Eq. (6.29), it is obvious that for sufficiently high LNA gain, the receiver NF is dominated by the NF of the LNA and the following stages have negligible contribution.

However, in practical applications high linearity is often required. Therefore, the proposed approach described in Fig. 7.10(b) as well as many radar systems omit the LNA in order to improve linearity at the expense of noise performance. In this case, the receiver noise figure is determined by the noise figure of the mixer. For a passive mixer the noise figure is approximately equal to the conversion loss [13]. However, the conversion loss of the receiver configuration in Fig. 7.10(b) is usually compensated for in the following IF amplifier stage. Thus, the overall NF of the receive path comprising a passive mixer is further increased by the NF of the IF amplifier, due to the conversion loss of the preceding stage. Therefore, the presented approach is applicable for systems that have moderate noise figure requirements.

7.2.2 Power-Amplifier-Mixer Circuit Design

The Power-Amplifier-Mixer (PAMIX) block, proposed in Fig. 7.10(b), has been realized on circuit level. The PAMIX schematic diagram is shown in Fig. 7.11.

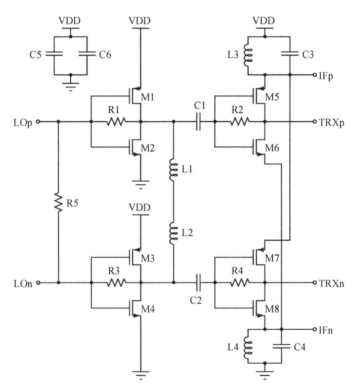

Fig. 7.11 Conceptual PAMIX schematic diagram.

7.2 Merged Power-Amplifier-Mixer Transceiver

The local oscillator (LO) signal, applied to the input pins LOp, LOn of the PA, is amplified and forwarded to the antenna through the pins TRXp and TRXn. The received signal is also applied to the pins TRXp and TRXn, whilst the down-converted signal is available at the pins IFp and IFn.

The mixing operation is realized by switching the transistors M5-M8 on and off by the strong amplified LO signal at the gates. During the positive LO half-wave the transistors M6, M7 are on, whilst during the negative half-wave the transistors M5, M8 are on. Thus, the RF signal at the drain is multiplied with square-wave-like switching characteristics of the transistors driven by a large LO signal. The resulting difference frequency is tapped and filtered at the sources of M5-M8. The mixing is thus realized similarly to a passive resistive mixer with the difference that the same transistor is also used for amplification. Thus, the DC drain-source voltage cannot be set to zero to keep the mixing transistor in the deep-triode region. However, due to the fact that the circuit functions as inverter, for "high" input the output is "low" with a minor delay. Therefore, during the on-state the mixing transistors are in the triode region and offer a low on-state resistance $R_{ds,on}$.

The shunt resonant circuits comprising L3, C3 and L4, C4 at the IFp and IFn nodes, respectively, function as IF band-pass filters. The DC bias is supplied through the inductors of these resonant circuits. At the RF and LO frequencies the capacitors C3, C4 offer very low impedance paths to ground. Thus, at these frequencies the sources of all four mixing transistors are shorted to AC ground.

At the IF frequency the resonant circuits exhibit high impedance. For this test circuit an IF frequency of 2 GHz has been chosen. The choice of the IF frequency depends on the target application. Higher IF frequencies offer the advantage of avoiding the flicker noise problem. Since one port of the device is connected to ground, the substrate parasitics of the inductors are partially shunted [14]. Therefore, the losses are reduced and higher Q-factors of the on-chip inductors are achieved.

The power amplifier is based on a two-stage pseudo-differential amplifier and very similar to the one described in section 7.1.1.

The large feedback resistors R1-R4 are used to stabilize the DC operating point, which is self-biased by appropriate choice of transistor widths. Resistor R5 is used to force the differential matching for measurement in a 50 Ω environment and would not be necessary if the circuit had been integrated with a VCO. The AC coupling capacitors C1, C2 are used to separate the self-biased operation points of the stages. The inductors L1, L2 are used to tune out the parasitic capacitance at the output of the first and at the input of the second stage.

Additionally, bypass capacitors have been added between VDD and GND to filter the supply voltage and to provide a low impedance path between VDD and GND at high frequencies.

7.2.3 PAMIX Measurements

The circuit has been realized in Infineon's C11N technology. The annotated micrograph is presented in Fig. 7.12. The chip area including pads is 0.28 mm².

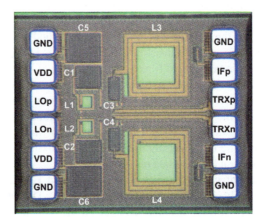

Fig. 7.12 Chip micrograph (chip size 0.59 mm × 0.47 mm).

The chip was thinned to 185 μm and mounted on a board for measurements. The circuit consumes 16 mA from a single 1.5 V supply. The mixer path has been characterized for RF input frequencies from 20 GHz to 27 GHz, whilst the IF was kept constant at 2 GHz. The input power at the PA input was 0 dBm and the input RF power −20 dBm. The measured and simulated conversion loss and SSB NF versus RF frequency are presented in Fig. 7.13 and Fig. 7.14, respectively.

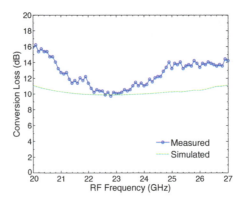

Fig. 7.13 Measured and simulated conversion gain versus RF frequency.

7.2 Merged Power-Amplifier-Mixer Transceiver

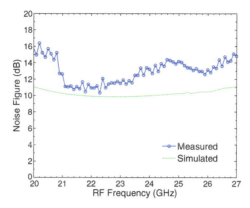

Fig. 7.14 Measured and simulated SSB NF versus RF frequency.

As can be seen, the mixer path achieves a conversion loss of 10 dB and a noise figure of 11.5 dB around the RF center frequency of 23 GHz. At the RF frequency of 24 GHz the circuit offers a sufficient performance of 11 dB conversion loss and of 12.9 dB noise figure, respectively.

The measured and simulated conversion loss versus IF frequency for a fixed RF frequency of 23 GHz is presented in Fig. 7.15. As can be observed, the mixer is well tuned to the IF frequency. The resonance in the vicinity of 1.2 GHz is due to the bondwire on the VDD pin with an inductance of ∼1.1 nH and the on-chip bypass capacitor of 15 pF.

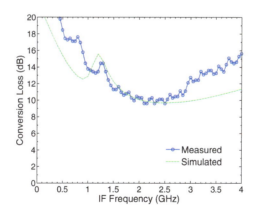

Fig. 7.15 Measured and simulated conversion gain versus IF frequency.

The measured input-referred 1dB compression point for an RF frequency of 23 GHz is 6 dBm. It has been observed in measurement that the input LO power of −2 dBm is sufficient to achieve an optimal conversion loss. The PAMIX concept offers the advantage of low required LO power compared to a standard passive mixer, since the input power is amplified before it is applied to the gates of the mixing transistors. Additionally, a very good port isolation has been observed over the whole frequency range. At an input frequency of 23 GHz the RF to IF isolation is 29 dB and the LO to IF isolation is 32.5 dB. The small-signal gain of the PA for an input power of −5 dBm is presented in Fig. 7.16. Gain values of 8 dB and 6.8 dB have been achieved at 21 GHz and 24 GHz, respectively.

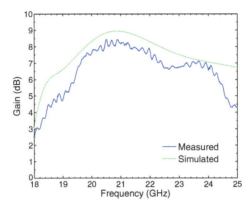

Fig. 7.16 Measured and simulated small-signal gain of the PA.

The large-signal performance of the power amplifier for the frequency of 21 GHz is presented in Fig. 7.17. The output-referred 1dB compression point of 2.5 dBm and saturation power P_{sat} of 7 dBm have been measured.

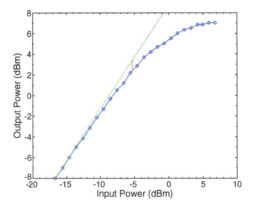

Fig. 7.17 Measured compression characteristics of the PA at 21 GHz.

7.2 Merged Power-Amplifier-Mixer Transceiver

It has been observed by comparing measurements of a stand-alone PA and PAMIX, that the power amplifier functionality remained unaffected.

7.2.4 Results Summary and Comparison

A novel concept of compact and cheap front-end circuits for monostatic radar applications has been presented. Due to the implementation of the mixer functionality directly at the output stage of the power amplifier, the need for an expensive duplexer has been resolved. Furthermore, a considerable chip area saving is possible due to omitting LNA and mixer blocks. The system-level considerations due to the absence of an LNA have been addressed.

The design of a merged power-amplifier-mixer block has been presented. The measurement results and comparison of the PAMIX parameters to the state-of-the-art power amplifiers and passive mixers are summarized in Table 7.2. The comparison is not straightforward, since the presented design is differential and measured on-board, whilst several publications use single-ended topologies and were measured on-wafer. However, it is clear that this work offers comparable performance and more functionality for smaller power consumption and chip area. The presented solution is attractive for low-cost mass-market radar applications.

Table 7.2 PAMIX measured performance summary and comparison

Parameter	[14]	[15]	[16]	[17]	This work
Topology	SE	SE	SE	SE	Diff
CMOS node (nm)	90	90	180	130	130
Freq (GHz)	26.5-30	9-31[1]	22.9-26	20-24	18-25
Size (mm^2)	0.12	1	1.26	0.35	0.27
Passive Mixer					
Loss (dB)	10.3	8-11[1]			10
SSB NF (dB)	11.4	NA			11.5
IP1dB (dBm)	NA	NA			6
LO-IF (dB)	22	22.5[1]			32.5
RF-IF (dB)	33	37.7[1]			29
LO Power (dBm)	0	9.7[1]			-2
Power Amplifier					
Gain (dB)			7	14.5	8
OP1dB (dBm)			11[2]	15	2.5
P_{sat} (dBm)			14.5[2]	16.8	7
P_{dc} (mW)			280	NA	16
V_{dc} (V)			2.8	3.6	1.5

[1] 1/2 LO source pumped.

[2] Measured on-wafer.

References

1. X. Guan and A. Hajimiri, "A 24-GHz CMOS front-end", *IEEE Journal of Solid-State Circuits*, vol. 39, pp. 368--373, February 2004.
2. Y.-H. Chen, H.-H. Hsieh, and L.-H. Hsieh, "A 24-GHz Receiver Frontend With an LO Signal Generator in 0.18-μm CMOS", *IEEE Transactions on Microwave Theory and Techniques*, vol. 56, pp. 1043--1051, May 2008.
3. A. Natarajan, A. Komijani, and A. Hajimiri, "A Fully Integrated 24-GHz Phased-Array Transmitter in CMOS", *IEEE Journal of Solid-State Circuits*, vol. 40, pp. 2502--2514, December 2005.
4. C. Cao, Y. Ding, X. Yang, J.-J. Lin, H.-T. Wu, A. K. Verma, J. Lin, F. Martin, and K. K. O, "A Fully Integrated 24-GHz Phased-Array Transmitter in CMOS", *IEEE Journal of Solid-State Circuits*, vol. 43, pp. 1394--1402, June 2008.
5. H. Krishnaswamy and H. Hashemi, "A Fully Integrated 24 GHz 4-channel Phased-Array Transceiver in 0.13 μm CMOS Based on a Variable-Phase Ring Oscillator and PLL Architecture", *in IEEE International Solid-State Circuits Conference (ISSCC)*, pp. 124--591, San Francisco, February 2008. IEEE.
6. Y. Cao, M. Tiebout, and V. Issakov, "A 24 GHz FMCW Radar Transmitter in 0.13 μm CMOS", *in Proc. of European Solid-State Circuits Conference (ESSCIRC)*, pp. 498--501, Edinburgh, UK, September 2008.
7. V. Issakov, M. Tiebout, K. Mertens, Y. Cao, A. Thiede, W. Simbürger, and L. Maurer, "A Compact Low-Power 24 GHz Transceiver for Radar Applications in 0.13 μm CMOS", *in IEEE Conference on Microwaves, Communications, Antennas and Electronic Systems (COM-CAS)*, pp. 1--5, Tel Aviv, Israel, November 2009.
8. M. Tiebout, "Low power, low phase noise, differentially tuned quadrature VCO-Design in standard CMOS", *IEEE Journal of Solid-State Circuits*, vol. 36, pp. 1018--1024, July 2001.
9. H.-D. Wohlmuth, D. Kehrer, and W. Simbürger, "A High Sensitivity Static 2:1 Frequency Divider up to 19 GHz in 120 nm CMOS", *in IEEE Radio Frequency Integrated Circuits (RFIC) Symposium Digest*, p. 231234, Seattle, USA, June 2002.
10. Agilent, "Application note 57-1: Fundamentals of RF and Microwave Noise Figure Measurements", http://www.home.agilent.com/, 2007.
11. D. Saunders, S. Bingham, G. Menon, D. Crockett, J. Tor, R. Mende, M. Behrens, N. Jain, A. Alexanian, and Rajanish, "A single-chip 24 GHz SiGe BiCMOS transceiver for FMCW automotive radars", *in IEEE Radio Frequency Integrated Circuits (RFIC) Symposium Digest*, pp. 459--462, Boston, USA, June 2009.
12. V. Issakov, M. Tiebout, H. Knapp, Y. Cao, and W. Simbürger, "Merged Power Amplifier and Mixer Circuit Topology for Radar Applications in CMOS", *in Proc. of European Solid-State Circuits Conference (ESSCIRC)*, pp. 300--303, Athens, Greece, September 2009.
13. D. N. Held and A. R. Kerr, "Conversion Loss and Noise of Microwave and Millimeter Wave Mixers: Part 1 -- Theory", *IEEE Transactions on Microwave Theory and Techniques*, vol. 26, pp. 49--55, February 1978.
14. F. Ellinger, "26.5-30-GHz Resistive Mixer in 90-nm VLSI SOI CMOS Technology With High Linearity for WLAN", *IEEE Transactions on Microwave Theory and Techniques*, vol. 53, pp. 2559--2565, August 2005.
15. M. Bao, H. Jacobsson, L. Aspemyr, G. Carchon, and X. Sun, "A 9-31-GHz Subharmonic Passive Mixer in 90-nm CMOS Technology", *IEEE Journal of Solid-State Circuits*, vol. 41, pp. 2257--2264, October 2006.
16. A. Komijani and A. Hajimiri, "A 24 GHz, +14.5 dBm Fully-Integrated Power Amplifier in 0.18 μm CMOS", *in Custom Integrated Circuits Conference (CICC)*, pp. 561--564, San Jose, USA, September 2005.
17. Y.-N. Jen, J.-H. Tsai, C.-T. Peng, and T.-W. Huang, "A 20 to 24 GHz +16.8 dBm Fully Integrated Power Amplifier Using 0.18 μm CMOS Process", *IEEE Microwave and Wireless Components Letters*, vol. 19, pp. 42--44, January 2009.

Chapter 8
Conclusions and Outlook

As mentioned throughout this work, there is a constantly growing demand for radar-based commercial sensors operating in the frequency bands around 24 GHz. The possible mass-market applications are numerous, and include door openers, theft alarms, motion sensors, sanitary equipment, and ground-speed measurement etc. Another very important market sector is automotive radar applications, such as e.g. lane change assistant and blind spot detection etc. However, one of the key factors influencing widespread market acceptance, apart from the technical features of the system, is the price of the module. As a result, there is tremendous pressure on keeping the hardware as inexpensive as possible.

One of the most challenging design blocks of a radar module is the RF transceiver front-end. Currently available integrated front-ends are typically realized using bipolar SiGe or GaAs compound semiconductor technologies that offer excellent high-frequency characteristics. This simplifies the circuit design efforts and provides higher performance margins. However, these technologies are not capable of high-level integration, particularly with digital processing blocks, and thus the radar modules comprise numerous discrete components assembled and interconnected on a board. This increases the overall integration, packaging and verification costs, which in turn increases the costs of the sensors.

In case of vehicular applications, current radar sensors are usually added only in the middle- and premium-class vehicles as an add-on option for an extra charge. The availability of a very low-priced vehicular radar sensor may provide the opportunity to incorporate numerous radar-based safety features as a standard fitment in any type of vehicle, including economy-class vehicles. Similarly, as seat belts, ABS or ESP became standard in almost every vehicle, the lane-change assistant or collision mitigation system may become standard features, as long as the module price is sufficiently low.

A considerable portion of the radar sensor electronics is based on digital circuitry. Therefore, using CMOS technology for realization of a highly integrated radar is particularly advantageous, since the highly integrated chip can include digital signal processing circuits for implementation of target detection algorithms.

V. Issakov, *Microwave Circuits for 24 GHz Automotive Radar in Silicon-based Technologies*, DOI 10.1007/978-3-642-13598-9_8, © Springer-Verlag Berlin Heidelberg 2010

However, the high-frequency characteristics of a standard 0.13 μm node are inferior compared to SiGe or GaAs alternatives. Thus, the realization of a 24 GHz radar front-end in this technology node is a challenge and its feasibility has been studied in this work. Furthermore, challenges related to modeling, characterization and ESD protection of microwave circuits are addressed.

This work proposes several numerical techniques for fast modeling of transistor parasitics as well as fitting of inductor S-parameters to an equivalent circuit model. Also, several techniques for de-embedding of differential or asymmetrical devices are proposed. Furthermore, characterization of differential devices using two-port network analyzers and baluns has been thoroughly analyzed and error estimation formulas have been derived.

The key receiver building blocks, LNA and mixer, have been realized in this work at 24 GHz in Infineon's 0.13 μm CMOS and 0.35 μm SiGe technologies in order to compare and contrast between the achievable performance and to estimate the necessity to use either technology for the aforementioned applications. All the presented circuits offer performance comparable with the state-of-the-art designs in terms of noise figure, gain, linearity, power consumption and area. The CMOS LNA is implemented using a unique circuit topology, whilst the SiGe LNA achieves the lowest reported noise figure at this frequency. Additionally, both circuits offer efficient ESD protection on the RF pads.

The major concern at higher flicker noise in a direct-conversion receiver front-end realization in CMOS has been addressed by considering passive mixers. However, also in bipolar circuits, a passive mixer realization can be useful, especially if a very high linearity is required. Therefore, the mixers have been realized in both CMOS and SiGe as active and as passive circuits. According to the author's knowledge both active mixers in CMOS and SiGe achieve the lowest reported noise figures. The inductorless passive mixer realization in CMOS and the bipolar passive mixer also provide a very wideband performance.

The system-relevant considerations of implementing passive mixers in a receiver have been thoroughly analyzed through the example design of two single-channel receivers comprising an LNA and a mixer implemented in CMOS. The active receiver has proved advantageous, provided it is possible to implement a continuous-wave radar with a frequency modulation scheme such that the IF is above 10 MHz.

For the same reason, further integration of the building blocks into integrated IQ receivers has been performed using active mixers. Again, both CMOS and SiGe receivers offer very low noise figure compared to literature. Additionally, the SiGe circuit offers good linearity and high ESD robustness on RF pins. Compared to CMOS, the SiGe receiver offers better performance at the expense of higher DC power consumption.

Furthermore, the receiver circuits have been analyzed over the extended automotive temperature range from -40 to 125 °C. The SiGe front-end offers much better temperature stability than the CMOS implementation.

Several techniques have been analyzed in this work that achieve the required very high ESD robustness without deterioration of the high-frequency characteristics.

8 Conclusions and Outlook

Finally, the CMOS receiver circuit was integrated with a transmitter consisting of a VCO, ILO, PA and divider. The circuit performance is comparable with that of a SiGe transceiver reported in literature. Furthermore, it compares favorably in terms of low noise figure.

The performance of the presented circuits shows that a standard 0.13 μm digital CMOS technology is suitable for realization of 24 GHz transceiver circuits. However, obtaining transmitter output power, receiver gain and linearity, as well as flicker noise comparable to what can be achieved in SiGe, is more cumbersome. Furthermore, performance stability under temperature variation, required for automotive applications, has proved to be disadvantageous in CMOS. Additionally, the presented SiGe circuits offer better parameters at the expense of higher DC power consumption. Therefore, the performance of the presented SiGe circuits is more advantageous for automotive applications requiring high robustness over temperature variations, whilst CMOS is more suitable for mass-market applications. Nevertheless, as soon as sufficient market volume is achieved in the automotive sector, a standard CMOS process may be considered as a compromise.

In order to further reduce the radar system costs, a novel transceiver integration concept has been developed in this work. It allows duplexers to be omitted and may thus considerably reduce module or chip size and costs.

The next step is to integrate the presented CMOS circuits with the base-band analog amplifiers and analog-to-digital converters (ADC) on the receiver side and with a synthesizer on the transmitter side. Furthermore, micro-controller circuits shall be added for digital signal processing, in order to fully profit by the integration capabilities of CMOS. On-chip integration of beam-steering antenna arrays is an interesting option. However, at 24 GHz the antennas are too large to be realized on-chip and the efficiency is too low due to substrate losses.

Apart from this, in order to be compatible with an external power supply, DC power conversion circuits, such as e.g. low dropout regulators or even DC-DC converter circuits, may be also integrated on chip. This is possible in CMOS technologies with sufficiently thick top metallization and high-voltage transistor options. In the case of a very high level system-on-chip integration, one of the main challenges for circuit design and layout is to provide sufficient cross-talk isolation between the power conversion and/or digital circuitry to sensitive RF and analog parts. This is particularly challenging for CMOS processes with highly-conductive substrates. An SOI CMOS technology would be advantageous for this purpose, but economically unsatisfactory.

Therefore, the following steps may focus on the integration of the presented circuits with analog, digital and power blocks in the standard digital 0.13 μm CMOS process. Along with an inexpensive packaging solution, a highly-integrated low-cost 24 GHz radar transceiver module can achieve high commercial success.

Appendix A
LFMCW Radar

For further insight on the linear frequency modulated (LFM) continuous wave (CW) radar, presented in section 2.3.2.1, this principle is described here analytically. Additionally, a LFMCW radar concept using ramps with different slopes for detection of multiple targets is described.

The sinusoidal transmitted signal can be written as

$$S_{TX} = A \cdot \cos{(\varphi_{TX}(t))}, \tag{A.1}$$

where A is the signal amplitude assumed to be constant and $\varphi_{TX}(t)$ is the instantaneous phase, related to the angular instantaneous transmitter frequency ω_{TX} and f_{TX}, respectively, as follows

$$\varphi_{TX}(t) = \int_0^t \omega_{TX}(t)\,\mathrm{d}t = 2\pi \int_0^t f_{TX}(t)\,\mathrm{d}t, \tag{A.2}$$

where t is the time since start of sweep and assuming $\varphi_{TX} = 0$ at time $t = 0$. For a CW radar operating at a single frequency f_{TX} is constant and equal to the carrier frequency $f_{TX}(t) = f_0$. It is assumed for simplicity that the phase at time $t = 0$ is $\varphi_{TX}(0) = 0$. For a linearly modulated radar a single rising frequency slope, as e.g. in the time interval $[0; T_m/2]$ in Fig. 2.3, can be expressed as follows

$$f_{TX}(t) = f_0 + \frac{B}{T_m/2}t = f_0 + \frac{2B}{T_m}t. \tag{A.3}$$

Thus, the phase of the transmitter signal for the LFM modulation is given by substituting (A.3) into (A.2) as follows

$$\varphi_{TX}(t) = 2\pi \left(f_0 t + \frac{1}{2} \cdot \frac{2B}{T_m}t^2 \right)\bigg|_0^t = 2\pi \left(f_0 t + \frac{B}{T_m}t^2 \right). \tag{A.4}$$

V. Issakov, *Microwave Circuits for 24 GHz Automotive Radar in Silicon-based Technologies*, DOI 10.1007/978-3-642-13598-9_9, © Springer-Verlag Berlin Heidelberg 2010

The signal received from the target is delayed by the round-trip time τ and has a different amplitude B

$$S_{RX} = B \cdot \cos(\varphi_{RX}(t)) = B \cdot \cos(\varphi_{TX}(t - \tau)), \qquad (A.5)$$

where $\varphi_{RX}(t)$ is the instantaneous phase of the received signal. As mentioned previously, a mixer produces a signal at the instantaneous difference frequency, referred to as IF or beat signal, which can be expressed as follows

$$S_{IF} = C \cdot \cos(\varphi_{IF}(t)) = C \cdot \cos(\varphi_{TX}(t) - \varphi_{RX}(t)). \qquad (A.6)$$

Therefore, combining (A.5), (A.6) and substituting into (A.4) the phase of the down-converted signal is obtained

$$\varphi_{IF}(t) = \varphi_{TX}(t) - \varphi_{TX}(t - \tau) = 2\pi \left(f_0 \tau + \frac{2B}{T_m} t \tau - \frac{B}{T_m} \tau^2 \right). \qquad (A.7)$$

The last term in (A.7) is negligible since usually holds $\tau/T_m \ll 1$. Considering the case described in Fig. 2.3 of a target at distance R, approaching with a constant relative velocity v_r, the round-trip delay τ is given by

$$\tau = 2 \cdot \frac{R - v_r \cdot t}{c}, \qquad (A.8)$$

and the phase of the down-converted signal can be explicitly written as

$$\varphi_{IF}(t) \approx 2\pi \left[\frac{2R}{c} f_0 + \left(\frac{4B}{T_m} \cdot \frac{R}{c} - \frac{2v_r}{c} f_0 \right) t + \frac{4B}{T_m} \frac{v_r}{c} t^2 \right]. \qquad (A.9)$$

The last term is known as range-Doppler coupling and is again negligible. Thus, the generated intermediate frequency f_{IF} is obtained by differentiating the phase

$$f_{IF} = \frac{1}{2\pi} \frac{\partial \varphi_{IF}}{\partial t} = \frac{4B}{T_m} \cdot \frac{R}{c} - \frac{2v_r}{c} f_0. \qquad (A.10)$$

The first term is related to the range and identical to (2.9) corresponding to a stationary target at distance R, whilst the second term is related to the relative velocity and identical to the Doppler frequency shift in (2.7). Thus, equation (A.10) corresponds to f_{b1} in (2.10), describing the beat frequency level for the rising slope, also referred to as the *up-chirp*. For the falling slope, also referred to as the *down-chirp*, the derivation is very similar, with the minus sign in (A.3) leading to (2.11). For a receding target the minus sign in (A.8) is replaced by a plus sign and the resulting expressions for f_{b1} and f_{b2} are interchanged.

The beat frequency in (A.10) is a function of the range R and the relative velocity v_r. A very common method to represent this dependency is by means of a *range-velocity* diagram [1]. Measurement of a single up-chirp will result in a straight line, as can be expected from (A.10) by rearranging it for v_r to be a function of R and $f_{IF} = f_{b1}$

A LFMCW Radar

$$v_r = f(R, f_{b1}) = \frac{2B}{f_0 T_m} \cdot R - \frac{c}{2f_0} f_{b1}. \tag{A.11}$$

In order to have an unambiguous detection of the range and velocity of a single target, an additional down-chirp is required. The solution is then derived graphically by searching for the cross-section of the two lines in the diagram, as shown in Fig. A.1(a). However, a real automotive radar system usually has to detect more than one target. Unfortunately, the LFM method, using up-chirp and down-chirp with the same gradient, yields four cross-sections on the range-velocity diagram in presence of two targets. This results in ambiguity, since only two cross-sections can describe valid targets $T1$ and $T2$, as depicted in Fig. A.1(b).

Fig. A.1 Range-velocity diagram of LFM using two chirps.

Thus, a four chirp waveform is usually used in practical radar systems [2]. The ramps have different gradients, but may occupy the same bandwidth B, as shown in Fig. A.2(a). The resulting lines corresponding to all four chirps intersect at one point for a valid target, as described in Fig. A.2(b).

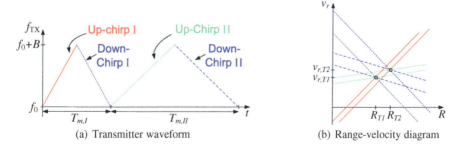

Fig. A.2 LFM using four slopes for unambiguous detection of two targets.

Implementation of four chirps is a trade-off between the probability of a false detection and the measurement time. This algorithm requires slow ramps in order to achieve sufficient velocity resolution. The slow ramps result in IF frequencies in the range of a few hundred kilohertz, which poses a challenge of front-end realization due to high flicker noise of MOS transistors, as discussed in section 3.3.1.

References

1. U. Lübbert, *Target Position Estimation with a Continuous Wave Radar Network*, Cuvillier Verlag, 2005.
2. V. Winkler, "Range Doppler detection for automotive FMCW radars", *in European Radar Conference (EuRAD)*, pp. 1445--1448, Munich, Germany, October 2007.

Appendix B
FSCW Radar

The frequency-stepped continuous wave (FSCW) technique is an additional approach that combines LFM with FSK [1]. The transmit waveform consists of two staircase modulated up-chirp signals A and B, having N steps in a bandwidth BW with an increment $f_{inc} = BW/(N-1)$. The signals are transmitted in an interwind manner, whilst A is used as a reference and B is shifted in frequency by f_{shift}, as shown in Fig. B.1(a).

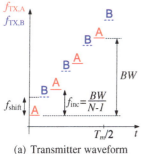
(a) Transmitter waveform

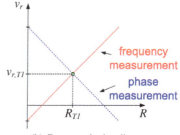
(b) Range-velocity diagram

Fig. B.1 FSCW principle.

The phase of the down-converted signal is evaluated for both signals. The phase difference $\Delta\varphi = \varphi_B - \varphi_A$ contains extended range and velocity information and is given by

$$\Delta\varphi = \frac{\pi}{N-1} \cdot \frac{v}{\Delta v} - 4\pi R \frac{f_{shift}}{c}, \tag{B.1}$$

where N is the number of frequency steps and Δv is the velocity resolution. Combining the two measurements of A and B and using the range-velocity diagram, unambiguous target range and velocity measurement is possible also for multiple targets, as shown in Fig. B.1(b). The main advantage of this method is that in several system implementations it is easier to generate frequency steps rather than a continuous sweep.

References

1. H. Rohling and M.-M. Meinecke, "Waveform design principles for automotive radar systems", *in CIE International Conference on Radar*, pp. 1--4, Beijing, China, October 2001.

Appendix C
Surface Charge Method

The following chapter briefly describes the implementation of the surface charge method (SCM) for calculation of external parasitic capacitances due to transistor multifinger layouts. Section C.1 provides the theoretical background of the surface charge method, whilst section C.2 presents meshing considerations.

C.1 Surface Charge Method Theory

The potential $\varphi(r_0)$ at any point r_0 in the vicinity of some conductors is given by the summation over the contributions of all very small surface areas dS with the charge densities $\sigma(r)$ around the points r on the conductors

$$\varphi(r_0) = \frac{1}{4\pi\varepsilon_r\varepsilon_0} \iint_S \frac{\sigma(r)}{|\mathbf{r}-\mathbf{r_0}|} dS. \tag{C.1}$$

The conductor surfaces are divided into N very small area elements ΔS_i ($i = 1,\ldots,N$), as shown in Fig. C.1(a). The charge distribution over the ith unit area is assumed to be uniform

$$\sigma(r_i) = \sigma_i. \tag{C.2}$$

Thus, these area elements are replaced by N equivalent point charges

$$q_i = \sigma_i \cdot \Delta S_i, \tag{C.3}$$

positioned on the conductor surface at the centers of the corresponding unit areas $\mathbf{r}_i$ ($i = 1,\ldots,N$), as depicted in Fig. C.1(b).

Therefore, the integral in Eq. (C.1) is replaced by a summation over the interactions of equivalent point charges located at the centers of the very small discretization element areas

$$\varphi(r_0) = \frac{1}{4\pi\varepsilon_r\varepsilon_0} \sum_{i=1}^{N} \frac{q_i}{|\mathbf{r}_i-\mathbf{r_0}|}, \tag{C.4}$$

V. Issakov, *Microwave Circuits for 24 GHz Automotive Radar in Silicon-based Technologies*, DOI 10.1007/978-3-642-13598-9_11, © Springer-Verlag Berlin Heidelberg 2010

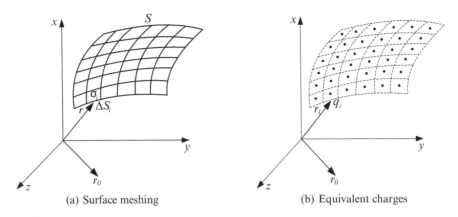

(a) Surface meshing (b) Equivalent charges

Fig. C.1 Surface charge method visualization.

where $\mathbf{r}_i$ is a center of the very small area ΔS_i, having a surface charge density of σ_i and equivalent charge of q_i.

When a voltage is applied between the conducting electrodes, the potentials at the discretization elements $\varphi(r_j) = \varphi_j$ define the boundary condition of the problem. Eq. (C.4) can be written for each point r_i and combined into a system of equations

$$X \cdot q = \varphi, \tag{C.5}$$

where X is known and related only to geometrical dimensions of the structure and discretization, φ is the vector of test potentials applied to the structure and q is the unknown vector of charges on the conducting bodies.

The system of equations in (C.5) can be explicitly written as follows

$$\begin{pmatrix} X_{11} & X_{12} & \ldots & X_{1N} \\ X_{21} & X_{22} & \ldots & X_{2N} \\ \ldots & \ldots & \ldots & \ldots \\ X_{N1} & X_{N2} & \ldots & X_{NN} \end{pmatrix} \cdot \begin{pmatrix} q_1 \\ q_2 \\ . \\ q_N \end{pmatrix} = \begin{pmatrix} \varphi_1 \\ \varphi_2 \\ . \\ \varphi_N \end{pmatrix}, \tag{C.6}$$

where X_{ij} ($i \neq j$) describes the interaction between equivalent point charges corresponding to two different small areas and X_{ii} describes the interaction within a single discretization unit area between the uniformly distributed surface charge to the equivalent point charge located at the center. The X_{ij} matrix entries are related to the distance between two charges and from (C.4) are given by

$$X_{ij} = \frac{1}{4\pi\varepsilon_r\varepsilon_0} \cdot \frac{1}{|\mathbf{r}_i - \mathbf{r}_j|}. \tag{C.7}$$

The derivation of the interaction inside a unit area X_{ii} is slightly more complicated. It requires the solution of a surface integral that contains a singularity. The solution thus depends on the shape of the discretization unit.

C.1 Surface Charge Method Theory

In case of a transistor multifinger layout, all the shapes are assumed to be rectangular. Therefore, the structure can be easily meshed by rectangular units. Thus, it is sufficient to analyse the self-interaction coefficient for a rectangular discretization element with an area of $\Delta S = a \cdot b$ and an uniform surface charge distribution of σ_0, as described in Fig. C.2.

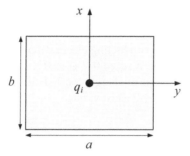

Fig. C.2 Rectangular unit area.

Due to symmetry considerations it is sufficient to solve only one quarter and multiply its contribution by four. Eq. (C.1) is used to calculate the potential at the center of the area, whilst the uniform charge distribution σ_0 is a constant

$$\frac{1}{|r_i - r_j|} \Rightarrow \frac{1}{\sqrt{x^2 + y^2}} \quad \text{and} \quad \varphi_{\text{center}} = 4 \frac{\sigma_0}{4\pi\varepsilon_r\varepsilon_0} \int_0^{a/2} \int_0^{b/2} \frac{1}{\sqrt{x^2 + y^2}} dxdy. \quad (C.8)$$

The indefinite integral of the function in (C.8) is given by

$$\int \frac{dx}{\sqrt{x^2 + y^2}} = \ln(x + \sqrt{x^2 + y^2}) = \sinh^{-1}\left(\frac{x}{y}\right). \quad (C.9)$$

The lower integration limit is now replaced by $\varepsilon \to 0$. Using the lower order terms of the Taylor expansion about $(x, y) = (0, 0)$ one obtains

$$\varphi_{\text{center}} = \frac{\sigma_0 ab}{2\pi\varepsilon_r\varepsilon_0} \left[\frac{1}{b} \ln\left(\frac{b}{a} + \sqrt{1 + \left(\frac{b}{a}\right)^2}\right) + \frac{1}{a} \ln\left(\frac{a}{b} + \sqrt{1 + \left(\frac{a}{b}\right)^2}\right) \right]. \quad (C.10)$$

Thus, the diagonal entries in X corresponding to the self-interaction of the surface charge with the equivalent charge in the center of an area are given by

$$X_{ii} = \frac{1}{2\pi\varepsilon_r\varepsilon_0} \left[\frac{1}{b} \ln\left(\frac{b}{a} + \sqrt{1 + \left(\frac{b}{a}\right)^2}\right) + \frac{1}{a} \ln\left(\frac{a}{b} + \sqrt{1 + \left(\frac{a}{b}\right)^2}\right) \right]. \quad (C.11)$$

Once the matrix X is calculated, the vector containing the values of the equivalent surface charges is obtained simply by

$$q = X^{-1} \cdot \varphi. \tag{C.12}$$

Now as the charge distribution for the applied test potential is determined, the static capacitance is calculated by considering the charge on the surface of one electrode. Assuming for example a system of two electrodes, the points $1, \ldots, k$ correspond to the first electrode and $k+1, \ldots, N$ to the second. The sample potential value of 1 V is set on the first conducting electrode $\varphi_i = 1$ ($i = 1, \ldots, k$), whist the potential of the other electrode is set to 0 V $\varphi_i = 0$ ($i = k+1, \ldots, N$). Thus, the capacitance is given by summation of charges located on one electrode q_i ($i = 1, \ldots, k$) and dividing the total charge by the test voltage

$$C = \frac{Q}{V} = \frac{\sum_{i=1}^{k} q_i}{1\,\text{V}}. \tag{C.13}$$

Obviously, for more complex system with several conductors the capacitances between the bodies are calculated similarly.

C.2 Meshing of the Multifinger Layout

The structure of a multifinger transistor layout, as for example depicted in Fig. 4.6, can be easily discretized using rectangular very small area elements, due to the fact that only rectangular faces are present in the structure. The basic meshed structure is a face of width W and length L, as shown in Fig. C.3.

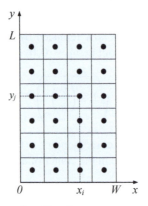

Fig. C.3 Discretization of a rectangular object face.

C.2 Meshing of the Multifinger Layout

The structure is divided into n_x units horizontally and n_y units vertically. Thus, the coordinates of the charge at the ith horizontal and j^{th} vertical unit area are simply given by

$$\{x_i, y_j\} = \left\{ \frac{W}{n_x} \left(i - \frac{1}{2} \right), \frac{W}{n_y} \left(j - \frac{1}{2} \right) \right\}. \tag{C.14}$$

This way the coordinates of all the equivalent surface charges are calculated and used in Eq. (C.7) and (C.11) to calculate the interaction matrix X and then the parasitic capacitance.

Rectangular meshing has an advantage of simplicity. Therefore, a procedure based on the surface charge method can be parameterized for a typical multifinger transistor layout and implemented as code in any programming environment, such as e.g. Excel, Matlab or C++. This can provide an accurate estimation of the parasitic capacitance due to transistor finger metallization within seconds without running a field solver or parasitic extraction.

Appendix D
Measurement of Active Circuits

This work presents numerous measurement results of active components, mainly LNAs, mixers, and receivers. The underlying measurement principles are described in this chapter. Section D.1 presents a brief overview about the measurement techniques used in this work. Since the majority of the presented circuits implement differential signaling, differential measurement setups are considered in the following sections. Section D.2 describes LNA measurements, whilst section D.3 presents the basic mixer and receiver measurements.

D.1 Measurement Techniques

The circuits presented in this work have been characterized through either on-wafer or on-board measurements. In the first case the pads of the DUTs have been directly contacted by on-wafer high-frequency probes. In the second case the chips were mounted on a printed circuit board (PCB) and the signals were accessed either through coaxial connectors or by contacting on-board lines, which are in turn connected to the pads of the DUT, using high-frequency probes.

The on-wafer measurements offer the advantage of high accuracy and good repeatability. Additionally, a large number of components can be easily verified on-wafer by an automated measurement system, in case that statistical distribution of the performance on a wafer is of interest. On-board measurement of hundreds of chips is not feasible due to material costs and assembly time of the test boards. However, constructing a corresponding setup using probes is very cumbersome and costly, particularly in the case that numerous pins have to be contacted simultaneously. Probes are available in certain standard pitches between the tips, as e.g. 100 μm, 150 μm or 200 μm and in various contact configurations as e.g. GSG, GSSG or GSGSG, where G signifies a ground tip and S signifies a signal tip. Obviously, the available probe pitches and contact configurations have to be known and taken into consideration already in the physical layout in order to secure the devices that can be contacted afterwards. In several cases when the pad density is too high, it

V. Issakov, *Microwave Circuits for 24 GHz Automotive Radar in Silicon-based Technologies*, DOI 10.1007/978-3-642-13598-9_12, © Springer-Verlag Berlin Heidelberg 2010

is not possible to position the probes accurately due to mechanical constraints. Furthermore, the probe tips are very delicate and require careful handling. An example of a spiral inductor measured on-wafer using two GSG Cascade Microtech Infinity probes is shown in Fig. D.1.

Fig. D.1 On-wafer measurement of a spiral inductor.

An example of a test board having coaxial connectors and used for on-board IQ receiver characterization is presented in Fig. D.2(a). The board implements hybrid ring couplers for single-ended to differential conversion at the RF and LO ports. The bonded bare receiver chip is shown magnified in Fig. D.2(b).

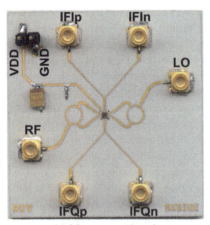
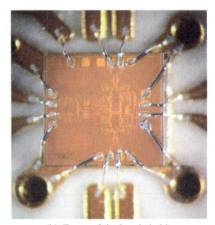

(a) Measurement board (b) Zoom of the bonded chip

Fig. D.2 Test board with baluns for coaxial on-board receiver measurement.

D.1 Measurement Techniques

For on-board measurements either a bare die or a packaged chip is mounted on a test board. At the frequency of 24 GHz standard inexpensive packages as e.g. TSLP or VQFP are applicable [1]. At even higher frequencies advanced packages as wafer level Ball Grid Array (BGA) can be considered [2]. On a board the signals are guided over microstrip or coplanar lines to the coaxial connectors. The main advantage of the on-board measurement approach is the flexibility of connections that allows measurement of complex chips with high pin-count. Furthermore, it enables realization of compact additional components as e.g. baluns, bypass capacitors or peripheral circuitry. However, the main disadvantage is the lower measurement accuracy and repeatability due to manual soldering and bonding. In order to achieve higher accuracy, a test board can be designed with probe launches instead of coaxial connector soldering pads and high-frequency probes for on-board measurements can be used.

During S-parameter measurements using a VNA, it is often not possible to set the reference planes directly at the ports of the measured DUT. Thus, structures between the calibrated reference planes and the DUT compose an error box. The impact of an error box is removed from the measured results by de-embedding. De-embedding is particularly critical for on-board coaxial measurements, since reference planes can be set by calibration only at the connector edge. Therefore, the on-board transition between the connector and the pads of the chip represents an error box that has to be de-embedded. A direct calibration, if possible in the particular case, would require manufacturing of test boards, test chips or test packages with well defined reference standards. This results in high manufacturing costs and complexity.

A transition from a bare chip to board using bondwires at 24 GHz is a challenge, since bondwires behave like inductors with very high quality factors. The inductance value is related to the physical length approximately as $1\,\mathrm{nH/mm}$. A bondwire, having a typical length of $300\,\mu\mathrm{m}$ and thus inductance of 0.3 nH, might detune the center frequency of a resonant circuit or affect matching. De-embedding of bondwires is very difficult. Therefore, the chip-to-board transition has to be modeled very carefully and considered during design stage. Additionally, all the chips in this work are thinned in order to reduce the bondwire length.

An additional challenge is the characterization of differential devices. When measured using a two-port VNA or any single-ended equipment, as rather usual for cost reasons, either on chip or external baluns become necessary. De-embedding accuracy of this approach has been addressed in detail in section 5.2.

Independent of the measurement type, on-wafer or on-board, the characterization techniques described in the following sections remain the same. In this work only the S-parameters of low noise amplifiers have been characterized both on-wafer and on-board, whilst measurements of other LNA parameters and of another active circuits have been performed on-board. Therefore, the LNA S-parameter section includes block diagrams of both measurement types, whilst the other block diagrams focus only on the on-board measurements.

D.2 LNA Characterization

D.2.1 S-parameter Measurement

The main small-signal characteristics of an LNA are determined from S-parameters that can be measured using a network analyzer. Particularly interesting properties are the gain, port matching and reverse isolation. Conceptual simplified block diagrams of an S-parameter measurement setup using a two-port VNA and a four-port VNA are presented in Fig. D.3(a) and Fig. D.3(b), respectively.

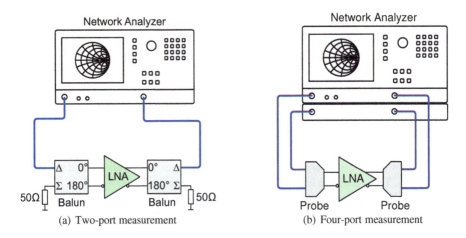

Fig. D.3 Conceptual block diagrams of S-parameter measurement setups.

Initially, the VNA is calibrated in the required frequency range using either coaxial standards for on-board measurements or using a calibration substrate for on-wafer measurements. Then S-parameters are obtained directly and can be further analyzed. All the two-port measurements of active devices in this work have been performed using HP's network analyzer 8510B with an S-parameter extension 8515B up to 26.5 GHz. The four-port S-parameter measurements have been performed using Agilent's network analyzer 8364A with the Z5623A multiport S-parameter test set extension up to 50 GHz.

D.2.2 Noise Figure Measurement

The noise figure of any device can be measured in several ways: either directly using the *noise floor* method, using the *signal generator* method or using the widely used *Y-factor method*. The latter advanced method requires a noise source that is

D.2 LNA Characterization

switched on and off for "hot" and "cold" measurements. These sources are usually specified by their excess noise ratio (ENR), which is defined as the ratio of noise power delivered to a 50 Ω load in on and off states. The ratio P_{hot}/P_{cold} is also referred to as the Y-factor. A very good overview of the noise figure measurement techniques is provided in [3].

The Y-factor method is also implemented for the operation of the noise figure meter (NFM) equipment. The noise figure measurements in this work are performed using HP's 8970B NFM that measures the noise figure and gain of the device. For LNA measurements the 8971C noise figure test set up to 26.5 GHz is required in order to down-convert the input signal to the range of the noise figure meter that operates up to 1600 MHz. A typical noise figure measurement setup for a differential LNA is depicted in Fig. D.4.

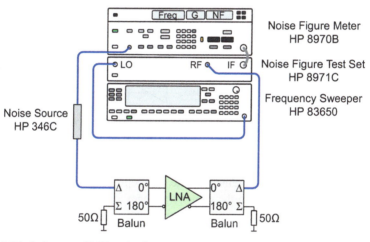

Fig. D.4 Block diagram of LNA noise figure measurement setup.

Most commercially available NFMs have integrated calibration procedures to account for the losses and noise figure of the setup. However, in case of the on-board measurement additional components as on-board baluns, traces and bondwires leading to the DUT ports have to be de-embedded. This is a non-trivial matter, since the effects of baluns on the noise measurements are difficult to de-embed. A technique for de-embedding noise figure of differential circuits using baluns and an extended Friis equation are proposed in [4]. However, the required two-port arrangements are difficult to realize in this case due to the bondwire transitions. Thus, in a frequency region, at which baluns offer negligible common-mode at the differential port, the noise figure can be de-embedded using the *Insertion Loss* technique [5] and the classical Friis equation [6]. Another advanced technique for de-embedding noise figure of a differential amplifier has been published recently [7]. This method is particularly useful for an on-wafer LNA noise figure measurement since it does not require external baluns.

D.2.3 Linearity Measurement

The linearity of an LNA is typically defined by the 1dB compression point (P1dB) and the third-order intercept point (IP3). These are either referred to the input or to the output of an LNA and thus denoted as IP1dB, IIP3 or OP1dB, OIP3, respectively. Usually, the IP3 is about 9 dB higher than the P1dB [8].

A typical setup for measurement of the 1dB compression point is presented in Fig. D.5. A signal, at the center frequency of the LNA, is applied to the input via baluns. The power level of the signal is increased gradually and the output power is observed on the spectrum analyzer (SPA). From the difference of these values and considering the setup losses, the gain is obtained. This gain is plotted versus the RF power and a point is determined at which the gain drops by 1dB.

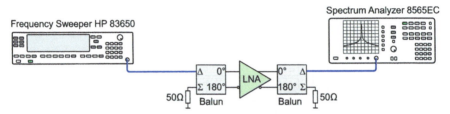

Fig. D.5 Block diagram of LNA 1dB compression measurement setup.

A typical setup for measurement of the intercept point is depicted in Fig. D.6. Two signals, usually close to the center frequency of the LNA e.g. 24 GHz and 24.001 GHz are applied to the input through baluns. The power level of both signals is increased gradually and the output power is observed on the SPA. In this case the intermodulation components are of interest, which appear at 23.999 GHz and 24.002 GHz. Thus, the power levels of the first harmonics and of the third order intermodulation products are plotted versus the input power. Lines with slope of 1 and 3 are positioned on the measured curves at low input power levels and extrapolated. Interception of these lines provides the IIP3 [8].

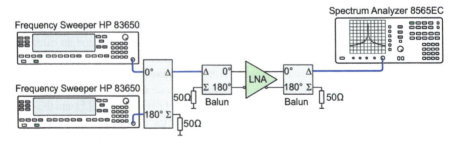

Fig. D.6 Block diagram of LNA intermodulation measurement setup.

D.3 Mixer and Receiver Characterization

The following measurements are applicable for mixers and receivers, since these devices perform frequency conversion and have similar characteristic parameters.

D.3.1 Conversion Gain Measurement

Receiver characterization can be performed using two signal sources and a spectrum analyzer. The sources are applied to the RF and at the LO ports of the receiver. The resulting signal at the difference frequency is measured at the IF port. Since active Gilbert mixers in this work have high impedance loads, an additional buffer is required. The buffer used for the measurements in this work is described in detail in [9]. The signals at the output of the buffer are combined using a hybrid ring coupler operating at the corresponding IF frequency range.

The power level at the output IF frequency is subtracted from the power level at the input RF frequency to obtain the conversion gain. Again, the losses are considered using the *Insertion Loss* method. A typical measurement setup of the conversion gain is presented in Fig. D.7.

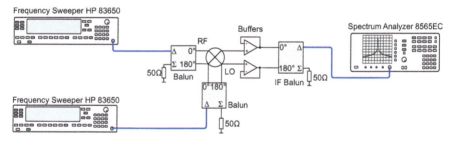

Fig. D.7 Block diagram of receiver conversion gain measurement setup.

Several sweeps are performed to obtain the conversion gain as function of different parameters such as RF frequency, IF frequency, LO power or RF power. In order to characterize a circuit over the RF frequency, RF and LO signals are swept, whilst the IF frequency is kept constant. For an IF frequency sweep, the RF frequency is swept, whilst the LO frequency remains constant.

D.3.2 Noise Figure Measurement

The noise figure of a mixer is obtained similarly as described in section D.2.2. In this case the down-conversion is performed by the DUT itself. Therefore, the measurement is performed only using a noise figure meter up to an IF frequency of 1600 MHz, a frequency sweeper and a noise source. Again, the losses of the setup have to be de-embedded using the Friis equation for noise figure of cascaded networks. It has been shown in [10] that the Friis equation is also applicable for a frequency converting chain, when the single sideband (SSB) noise figure is considered. A noise figure meter provides the double sideband (DSB) noise figure, which is related to the SSB noise figure simply by

$$\text{NF}_{\text{SSB}} \text{ (dB)} = \text{NF}_{\text{DSB}} \text{ (dB)} + 3 \text{ (dB)}, \tag{D.1}$$

provided the conversion gain in the RF and image bands is the same [11]. In case that RF and LO frequencies are very close to each other, as in case of a low-IF or a direct down-conversion receiver, only the double sideband noise figure can be measured. A typical measurement setup for the noise figure characterization of a receiver is presented in Fig. D.8. Typically, the noise figure is measured over RF or IF frequency. Additionally, a sweep over LO power is performed and the conversion gain and noise figure are plotted. This allows to estimate the lowest LO power at which the mixer or receiver provides the optimal performance.

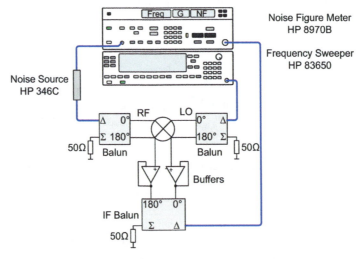

Fig. D.8 Block diagram of receiver noise figure measurement setup.

D.3.3 Linearity Measurement

Linearity characterization of a receiver is very similar to that of an LNA, described in section D.2.3. The 1dB compression point is obtained from the same setup as in case of the conversion gain measurement, described in Fig. D.7. In this case, the gain is swept over the RF power. The IP3 measurement setup is presented in Fig. D.9. In this case two frequencies are applied at the RF port. These components are mixed with an LO frequency and at higher power levels third order intermodulation harmonics are observed in the IF range. For example, for RF frequencies of e.g. 24.010 GHz and 24.011 GHz, an LO frequency of 24 GHz, the first harmonics are 10 MHz and 11 MHz, whilst the third order harmonics are 9 MHz and 12 MHz.

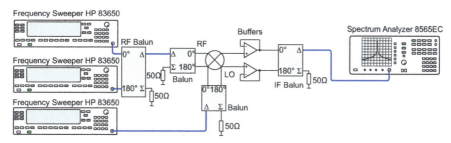

Fig. D.9 Block diagram of receiver IP3 measurement setup.

References

1. M. Engl, K. Pressel, J. Dangelmaier, H. Theuss, B. Eisener, W. Eurskens, H. Knapp, and W. Simbürger, "A 29 GHz Frequency Divider in a Miniaturized Leadless Flip-Chip Plastic Package", in *IEEE MTT-S International Microwave Symposium (IMS) Digest*, pp. 477--480, Fort Worth, USA, June 2004.
2. M. Wojnowski, M. Engl, B. Dehlink, G. Sommer, M. Brunnbauer, K. Pressel, and R. Weigel, "A 77 GHz SiGe mixer in an embedded wafer level BGA package", in *Proc. IEEE Electronic Components and Technology*, pp. 290--296, Lake Buena Vista, USA, May 2008.
3. Agilent, "Application note 57-1: Fundamentals of RF and Microwave Noise Figure Measurements", http://www.home.agilent.com/, 2007.
4. A. A. Abidi and J. C. Leete, "De-Embedding the Noise Figure of Differential Amplifiers", *IEEE Journal of Solid-State Circuits*, vol. 34, pp. 882--885, June 1999.
5. S. Belkin, "Differential Circuit Characterization with Two-Port S-Parameters", *IEEE Microwave Magazine*, vol. 7, pp. 86--99, December 2006.
6. D. Pozar, *Microwave Engineering*, Wiley, 2nd edition, 1998.
7. L. Belostotski and J. Haslett, "A technique for differential noise figure measurement with a noise figure analyzer", *IEEE Microwave Magazine*, vol. 10, pp. 158--161, February 2009.
8. F. Ellinger, *Radio Frequency Integrated Circuits and Technologies*, chapter 4, Springer, 2007.
9. W. Simbuerger, *Integrated Single-Chip Direct Conversion RF-Transceiver*, Dissertation, Institut für Nachrichten- und Hochfrequenztechnik der TU Wien, 1999.
10. B. Dehlink, H.-D. Wohlmuth, K. Aufinger, T. F. Meister, J. Böck, and A. L. Scholz, "A low-noise amplifier at 77 GHz in SiGe:C bipolar technology", in *IEEE Compound Semiconductor IC Symposium (CSICS) Digest*, pp. 287--290, Palm Springs, USA, November 2005.
11. T. H. Lee, *The Design of CMOS Radio Frequency Integrated Circuits*, Cambridge University Press, 1998.

Index

B7HF200 SiGe bipolar technology, *see* Technology
Back end of line (BEOL), 29
Balun, 48, 63, 87, 132, 153
Base
  Doping, 23
  Resistance, 23, 85, 96

C11N Si CMOS technology, *see* Technology
Coherent processing interval (CPI), 7

Deep trench isolation (DTI), *see* Trench isolation
Diode-triggered SCR (DTSCR), 148
Doppler effect, 8

Electrostatic discharge (ESD)
  Protection, 145
    Local clamping, 146
    Rail-based, 86, 146
    Transformer-based, 153
    Virtual ground, 147
  Testing standards, 145
  HBM, 130, 145
  Transmission line pulse (TLP) technique, 88
Equivalent isotropically radiated power (EIRP), 13
Excess noise ratio (ENR), 201

Frequency regulation
  Industrial, scientific and medical (ISM) band, 2, 13, 16, 167, 172
  Ultra-wide band, 13
Friis formula
  Noise figure of cascaded stages, 116
  Transmission, 6

Front end of line (FEOL), 20

Gate
  Oxide, 20, 22
    Breakdown, 150
  Resistance, 21, 79

Hybrid ring coupler, 69, 72, 87, 105, 136

Inductor
  Equivalent circuit, 34
  Quality factor, 30, 52, 82, 97, 175
Intermodulation free dynamic range (IMFDR), 118

Linearity
  1dB compression point, 8, 88, 116
  Characterization, 202, 204
  Input-referred third-order intercept point (IIP3), 88, 117
Low-noise amplifier (LNA)
  Bipolar, 83, 86
  Characterization, 200
  CMOS, 78, 86, 148, 155

Measurement
  Calibration, 48
    Line-Reflect-Reflect-Match (LRRM), 48
    Multiline TRL, 58
    Multimode TRL, 54
    Short-Open-Load-Thru (SOLT), 48, 60, 71
    Thru-Reflect-Line (TRL), 48
  De-embedding, 47
    Insertion Loss, 63, 87, 201
    Open-Short, 48, 137
    Thru, 49

208   Index

Thru-Line (TL), 48, 60
Thru-Reflect-Line (TRL), 48, 60
Metal-insulator-metal (MIM) capacitor, 22, 25, 158
Mixer
  Active
    Bipolar, 95
    CMOS, 93
  Characterization, 203
  Passive
    Bipolar, 105
    CMOS, 102
Monopulse principle, 11
  Amplitude-comparison monopulse, 12
  Phase-comparison monopulse, 12

Noise
  Factor, 7, 79, 116
  Figure (NF), 8, 27, 79, 85, 116
    Characterization - noise floor, 171, 200
    Characterization - signal generator, 200
    Characterization - Y-factor, 200
    Double sideband (DSB), 204
    Single sideband (SSB), 8, 116, 204
  Flicker, 14, 92, 111, 114, 116
  Floor, 7
  Phase, 170
  Resistance, 79, 85
  Thermal, 7, 79

Oscillator
  Injection-locked (ILO), 167
  Voltage-controlled (VCO), 167

Polyphase filter, 123, 144
Power amplifier, 168
Power amplifier mixer (PAMIX), 174
Power Combiner
  180° on-chip, 132
  90° on-chip, 134
Power Splitter, *see* Power Combiner

Radar
  Bistatic, 6
  Continuous wave (CW), 8, 13
  Frequency-modulated continuous wave (FMCW), 9, 185
  Frequency-stepped continuous wave (FSCW), 9, 189

Long-range radar (LRR), 17
Mid-range radar (MRR), 16
Monostatic, 6, 165, 173
Principle, 5
Short-range radar (SRR), 16
Ultra-wideband (UWB), 16
Radar cross-section (RCS), 6
Rat-race coupler, *see* Hybrid ring coupler
Receiver
  Characterization, 203
  Heterodyne, 15
  Homodyne (direct down-conversion), 14
  In-Phase/Quadrature(IQ), 122, 124
  Low-IF, 16
  Single-channel, 112
    Active, 113
    Passive, 113

Shallow trench isolation (STI), *see* Trench isolation
Signal to noise ratio (SNR), 7
Skin effect, 30, 34
Spurious free dynamic range (SFDR), *see* Intermodulation free dynamic range (IMFDR)
Surface Charge Method (SCM), 42, 191

Technology
  Si CMOS, 19, 26, 77
  SiGe bipolar, 23, 26, 77
Transformer
  Equivalent circuit, 153
  Interleaved, 133
  Stacked, 60
Transistor
  Bipolar homojunction transistor (BJT), 23, 77
  Heterojunction bipolar transistor (HBT), 23, 24, 26, 77
  Metal-oxide-semiconductor field-effect transistor (MOSFET), 20, 26, 77
  Transconductance g_m, 27, 83, 85, 91, 120
  Transit frequency f_T, 26, 91
Transmit/receive (T/R) switch, 6, 165
Trench isolation
  Deep (DTI), 23
  Shallow (STI), 20

Lightning Source UK Ltd.
Milton Keynes UK
UKOW06n2023210316

270575UK00001B/30/P